Honoring the Veterans of Licking County, Ohio

An llustrated history of Licking County's military heritage

Sixth Street Park in Newark

Honoring the Veterans of Licking County, Ohio

An llustrated History of Licking County's Military Heritage

Aaron J. Keirns & Nathan J. Keirns

Little River Publishing
Howard, Ohio

★

For those who served.

Other books by Aaron J. Keirns

Black Hand Gorge ~ A Journey Through Time

The illustrated history of a spectacular river gorge in Licking County, Ohio. Includes chapters on the Ohio & Erie Canal, railroad, electric interurban railway and more. Contains over 150 photographs & illustrations.

Ohio's Airship Disaster

The illustrated story of the tragic crash of America's first rigid dirigible, the USS Shenandoah. In September of 1925, the giant Navy airship was ripped apart by a storm over Noble County, Ohio. The ship's captain and 13 of his men died in the crash. Contains over 100 photographs and illustrations.

Statues on the Hill

The illustrated story of Baughman Memorial Park where ten larger-than-life statues of U.S. presidents and Civil War generals stood forgotten on a hilltop in Muskingum County, Ohio. Carved out of native sandstone by an undertaker, these statues tell a remarkable story of one man's vision. Contains over 100 photographs & illustrations.

To order this book or any of the books listed above:
Send a check or money order for $25 per book (which includes sales tax & shipping) to:

Little River Publishing
23700 Cornish Road
Howard, OH 43028

Please be sure to indicate the title(s) of the book(s) you are ordering and the mailing address where you would like them shipped (U.S. addresses only).

For bookstore orders, please contact the publisher.

Honoring The Veterans of Licking County, Ohio

ISBN 0-9647800-6-2

Little River
PUBLISHING
HOWARD, OHIO

Contents

Left: Stone monument in the veterans section of Sixth Street Park, Newark, Ohio.

Perryton Cemetery

ACKNOWLEDGEMENTS

The authors wish to thank the many individuals and organizations that helped provide information and inspiration for this book.

We appreciate your help!

Beverly Adzic
Vince Becker Jr.
Ken Blaney
Dan Fleming
Sondra Brockway Gartner
C. Franklin Gibson
Philip Gregg
Gloria Hoover
Emily Larson
Linda Leffel
Walter Ludwig
Edward McDonald
Terry Mooney
Mary C. Oliver
Connie L. Rutter
James & Janet Seymour
Larry Stevens

American Legion
AMVETS
Granville Public Library
Licking County Genealogical Society
Licking County Historical Alliance
Licking County Historical Society
Licking County Library
Licking County Veterans' Service Commission
National Heisey Museum
Newark Advocate
The Ohio Historical Society
The Works: Ohio Center for History, Art & Technology
Veterans of Foreign Wars
Vietnam Veterans of America

And, we thank our families for their patience and support while we spent endless evenings and weekends on this project.

Cedar Hill Cemetery, Newark Ohio

This book is a tribute to military veterans, particularly those with ties to Licking County, Ohio.

The book is not intended to be a scholarly or comprehensive history of wars, military units or individuals. There are already plenty of good books available on those subjects. Our goal in creating this book is simply to bring together, in one place, local information that is currently scattered across the pages of a variety of books, documents, government records, markers and other places.

The idea is to create a single, easily accessible source of basic information about Licking County's involvement in America's wars and conflicts, and to provide a visual record of the county's many statues, plaques and memorials dedicated to veterans.

We hope this little volume will be a useful resource for teachers, history buffs and students of all ages.

Aaron J. Keirns
Vietnam-era Veteran

Nathan J. Keirns

You can help us make this book better.

This book is meant to be a record for the community. We realize there are likely some errors and omissions in this first edition. We will consider republishing the if needed in order to make corrections or incorporate additional information that we may receive from you, our readers.

The publisher encourages readers to point out any errors so we can correct them. We would also like to add information that is missing or incomplete. If you have *reliable* information, documents, photographs or stories that you think would add to the usefulness of this book, please let us know. We will consider all submissions for inclusion. Please, do not send us original documents. Send copies that don't need to be returned.

If you know of veterans monuments of any kind in Licking County that we missed, we'd like to know about them. *Please note, we are not looking for individual headstones or grave markers as there would be far too many to show. We are looking for public monuments, statues, plaques or other memorials honoring veterans' groups, military units or veterans in general.*

To contact the publisher, write to:
Little River Publishing, 23700 Cornish Rd., Howard, Ohio 43028
Or send an e-mail to: **info@littleriverpublishing.com**

From Collection
Jas. T. Haynes
REUNION OLD FORT
NEWARK
76 Regt
O.V.I.
JULY 22 1890
Reunion at NEWARK
76th O.V.I. Fair Grounds July 22-90
From Collection of James T. Haynes Sr.

OUR MILITARY HERITAGE

It would be impossible to document all the veterans who have called Licking County their home over the last 200 years.

Many of the early explorers and settlers who came to the area were military veterans before they got here. They had served in the Revolutionary War, The War of 1812 or other conflicts.

Licking County was officially formed in 1808. Since that time, countless men and women have answered the call to duty. The contributions and sacrifices made by this ever-growing army of citizens are immeasurable.

MILITARY DISTRICT

Licking County's military heritage goes all the way back to the land itself. A large section of central Ohio, including Licking County, was once part of The United States Military District. This district contained over two and a half million acres set aside for veterans of the Revolutionary War (or their widows or heirs) as partial compensation for their service.

The size of each veteran's land grant was determined by his rank: Major General – 1,100 acres, Brigadier General – 850, Colonel – 500, Lieutenant Colonel – 450, Major – 400, Captain – 300, Lieutenant – 200, Ensign – 150, non-commissioned officer or soldier – 100. Some veterans settled on the land they were given but many had no desire to live this far "west," and sold their land without ever seeing it.

THE REFUGEE TRACT

Along the south edge of the Military District, a narrow strip of land containing 103,527 acres was set aside as the "Refugee Tract,"

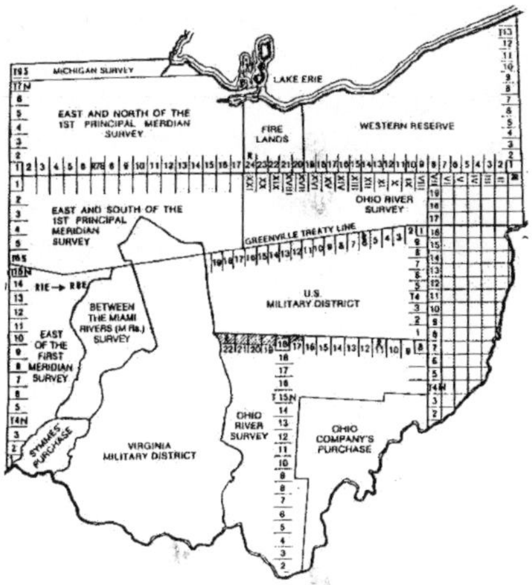

Above: Ohio land survey map showing United States Military District encompassing area now occupied by Licking County.

(Source: Auditor, State of Ohio)

Opposite: Civil War veterans from the 76th Ohio Volunteer Infantry gather for a reunion on July 22, 1890 on the bank of the Great Circle Earthworks at Moundbuilders Park.

(Photo: Licking County Historical Society)

Photo on page 8: Canon in Cedar Hill Cemetery, Newark Ohio. End of barrel is stamped with the following:

"S.C.L. No. 852 P.I. Co. 1865 816 lbs."

Above: John Lincoln Clem, the "Drummer Boy of Chickamauga," was born in Newark, Ohio in 1851. This image is from *Harper's Weekly*, Feb. 6, 1864, illustrating an article entitled: Sergeant John Clem, Our Youngest Soldier."

Above: William Burnham Woods was a Newark native who served as a general in the Civil War and later became a Supreme Court Justice.

part of which was in Licking County. The purpose of the tract was to set aside land for refugees from Canada and Nova Scotia (or their widows or heirs) who had fled those British provinces in order to assist the colonial cause during the Revolutionary War. When these people left Canada and Nova Scotia, their land and possessions there were confiscated by the British government. The Refugee Tract was intended to compensate the refugees for their loss of property. The Refugee grants ranged from 320 to 2,240 acres depending on the individual's loss.

Only 67 claimants came forward, claiming a total of 58,080 acres. The remaining land reverted to the government and was sold.

Refugee Road, which runs from Hebron to Columbus takes its name from this tract.

Early Military Groups

One account of early Newark tells of a highly anticipated General Muster that was held in the fall of each year. Loosely formed companies of men such as the "Hog Run Rifles," would come into town (from the vicinity of Hog Run) looking to settle old grievances and to partake in generous amounts of gingerbread, cider, beer and whiskey. These gatherings usually involved contests of physical strength and agility that often deteriorated into raucous political debate and fighting.

First Formal Military Company

Organized in 1856, the "Newark Guards of Light Artillery of Ohio Militia," was Newark's first formal military company. The company was armed with muskets and a two-pounder light field pieced (photos at right). A military band was also part of the organization. The militia disbanded when the Civil War broke out and many of the members left to join the Union Army.

Wide-awakes

During the presidential campaign of 1860, semi-military organizations known as "Wide-awakes" formed all over the country, including Licking County. The groups were formed to maintain peace during political rallies and to make sure the results of voting were protected. They often paraded around political gatherings armed, wearing capes and carrying torches.

The Squirrel Hunters

Today our soldiers are furnished with uniforms, high-tech weapons, transportation and anything else needed for their military duty.

But during the Civil War things were not so standardized. Men known as the "Squirrel Hunters," answered the call of duty on the spur of the moment, and they brought their own hunting rifles.

When the city of Cincinnati was threatened by Rebels in September 1862, Ohio's governor called upon the state's citizens to immediately send squads of armed men for a short term of service. Licking County furnished 404 men from all walks of life in answer to the request, as described by E.M.P. Brister in the *Centennial History of Newark and Licking County:"*

"*...church and fire bells rang, mounted men galloped through neighborhoods to spread the alarm; there was a hasty cleaning of rifles and moulding of bullets, and filling of powder horns, and mustering at the villages; and every city-bound train ran burdened with the gathering host.*"

The Cincinnati skirmish was short-lived and the invaders "retired" after only a few days. Each Squirrel Hunter was eventually given a month's militia pay ($13) for his service.

Above & Below: Members of the Newark Guards of Light Artillery of Ohio Militia in 1858, mustered on Newark's Courthouse Square in front of the Preston House, which later became the Lansing House.

(Photo: Licking County Historical Society)

VOLUNTEERS
FOR THE
76th Ohio Infantry.
A RECRUITING RENDEZVOUS for this old and tried regiment has been opened in Newark, Ohio. The most liberal bounties will be paid for able bodied men, between the ages of 18 and 45. Office on the corner of Main street and the Square, over Patton's store.
W. B. WOODS, Lieut. Col.,
JNO. J. METZGAR, Lieut.,
Newark, Sept. 4, '63—tf6 [Recruiting Officers.

Above: An advertisement in the Advocate, September 4, 1863.

Below: A 1902 map showing the Camp McKinley rifle range at the Octagon Mound (now Moundbuilders Country Club golf course). The "Rifle Range" indicated by the strip of parallel lines, was 1,000 yards long.

(Business Map of the City of Newark and Suburbs, by Chas. H. Wells & Co., 1902)

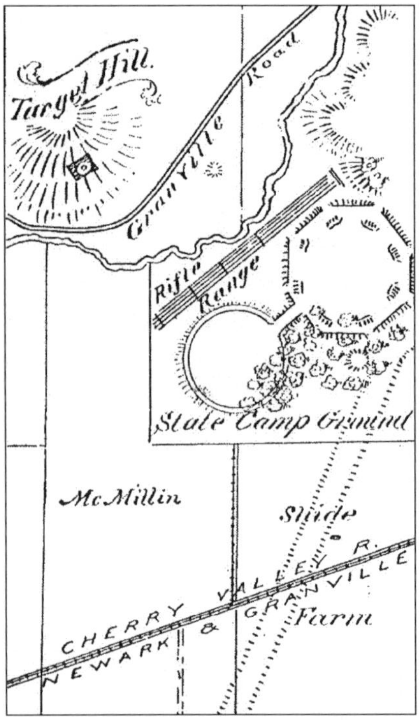

Right: Union Army veterans gathered on the steps of the Licking County Courthouse around 1896.

(Photo: Brockway, Images of America – Newark)

The 76th Ohio Volunteer Infantry

Known as the "Licking County Regiment" or "Licking Volunteers," the 76th Ohio Volunteer Infantry (OVI) formed in the fall of 1861 at Camp John Sherman (now Moundbuilders Park) in Newark. Of the 10 companies making up the regiment, 7 were from Licking County. The regiment had 962 officers and enlisted men when it mustered in on February 9, 1862. A few days later, the regiment participated in the Battle of Fort Donelson, Tennessee.

The men of the 76th were in for a three-year enlistment. During that time, the regiment participated in many major battles, including the Battle of Shiloh, the Siege of Vicksburg, the Battle of Atlanta and Sherman's March to the Sea.

During its service, the regiment lost 9 officers and 82 enlisted men in battle and 5 officers and 265 enlisted men to disease.

Earthwork Encampments

In addition to being used as a golf course, amusement park and fairgrounds, Newark's prehistoric earthworks have also been commandeered by the military for use as training facilities.

The Great Circle Earthworks at Moundbuilders Park (where today South 21st Street meets Route 79) was once the site of a military encampment known as Camp John Sherman. The 76th Ohio Volunteer Infantry trained here in 1861-62. After the war, the 76th held reunions at the site.

The Octagon Mound (now the site of Moundbuilders Country Club golf course along 30th Street) also served as a military encampment originally called Camp Buckeye, later named Camp William McKinley. From 1892 to 1908 the Ohio State Militia (now called the National Guard) trained there.

Left: This painting by an unknown artist, dated 1903, depicts Camp McKinley at the Octagon Earthworks.

Veteran Reunions

Veterans of the Civil War seem to have had extremely strong and long lasting bonds of friendship between them. For years after the war communities across the north and south held regular reunions, sometimes on the same battlefields where they had once fought.

One annual reunion was held in August of 1900 in southeastern Licking County at Wickham's Grove near Toboso. A Zanesville newspaper described the scene:

"The annual reunion of the soldiers and sailors of the Civil War was held in Wickham's Grove near Black Hand on Thursday and was one of the most successful of this notable series. Between 5,000 and 6,000 people gathered at the assembly grounds to witness or participate in the exercises and to do honor to those veterans for whom no honor is too great."

The Great 'Re-Union' of 1878

A veterans reunion held at "The Old Fort" (Moundbuilders Park) in July of 1878 was one of the most memorable events in Licking County history. It attracted thousands of attendees.

Known as the "Great Re-Union," it marked the 14th anniversary of the death of General James B. McPherson who was killed in battle in 1864. In attendance were President Rutherford B. Hayes, Ohio Governor Richard M. Bishop, General William T. Sherman, General Phillip H. Sheridan and other well known dignitaries.

When President Hayes and General Sherman stepped up on the speakers' platform, the over-enthusiastic crowd surged-in causing the platform to come crashing down, which was followed by loud laughter and cheers. The Governor's Guards were then placed near the platform to prevent further crowding. The event concluded with a fireworks display that lasted an hour and a half.

"County's Last Civil War Veteran Succumbs"

The Advocate, June 5, 1939.

Frank D. French, 95, dies Sunday at His Home in Johnstown.

Military organizations in the county will join Wednesday and pay a last tribute to Frank D. French, last civil War veteran in Licking County, when burial is made with full military rites, under the auspices of Johnstown post, American Legion.

Mr. French, who observed his 95th birthday anniversary May 21st, died quietly Sunday morning at 11:45 o'clock in his home in Johnstown. He was able to speak to members of his family until an hour before death and then slept away.

Above: Frank D. French served four years with the 76th Ohio Volunteer Infantry and fought at Fort Donaldson, Pittsburgh, Corinth, Memphis, Vicksburg, Gettysburg, Chattanooga, Lookout Mountain, Missionary Ridge, Ringold and other battles. He was hit in the hip by a musket ball at Peachtree, Georgia and was discharged seven months later after spending time in a hospital near Stone Mountain.

Above: The old Memorial Building, which was located directly across North Second Street from where the Midland Theater now stands.

Note the arrows pointing to the statues of a soldier and sailor. These two statues are now located at the V.F.W on Forry Avenue in Newark

(Photo: Brister, Centennial History, 1909)

SOLDIERS & SAILORS MEMORIAL BUILDING

In 1891, a local G.A.R. group (Grand Army of the Republic) proposed the idea of building a community building to memorialize the soldiers and sailors who had died in the Civil War. A levy was passed to raise funds and construction began. The impressive stone structure opened in 1894.

The building housed the Auditorium Theater and was located on the northeast corner of the Courthouse Square. The Auditorium was considered one of the finest theaters in the Midwest. Ethel Barrymore, Ed Wynn, Gene Autry, Nelson Eddy and other legendary stars performed on its stage.

On a balcony above the front entrance, were life-size metal statues of a soldier and a sailor. In July, 1954, high winds blew the statues down. They were subsequently moved to their present location in front of V.F.W. Post 1060 on Forry Avenue.

In 1968 a fire destroyed the front facade of the Auditorium. Attempts were made to revitalize the building over the next few years but it was eventually torn down in 2002.

Left: These three statues in front of the V.F.W. on Forry Avenue in Newark were moved there from other locations. The Civil War soldier on the left and sailor on the right originally stood above the entrance of the Soldiers & Sailors Memorial Building and Auditorium Theater in Newark. The WWI soldier in the middle once stood in Sixth Street Park.

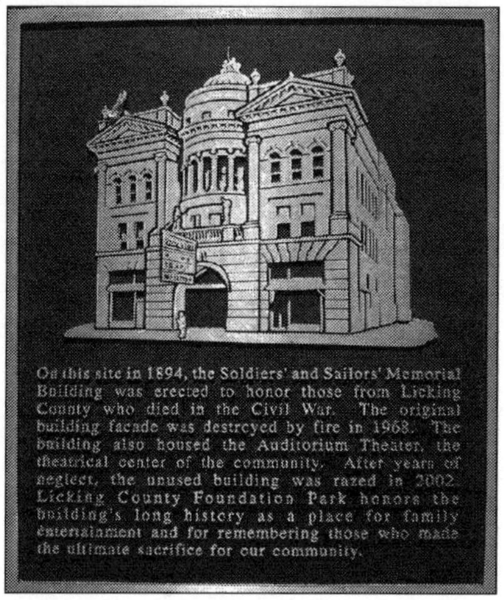

Above: Plaque now located at the site of the old Auditorium Theater in Foundation Park in downtown Newark.

Below: Faces of the three statues shown above left.

Women and War

Although their contributions are often overlooked in history books, women have always taken active roles during times of war. From providing food, clothing and medical care, to working in munitions factories and even joining in the fight, women have supplied critical support to troops throughout history.

Right: A group of Newark women who wore military-style uniforms and gave aid to soldiers in hospitals during the Civil War. The women (left to right) are: Helen King Spanger, Libbie Sparague Hamilton, Mary Warner Wilson and Mary (Mame) Ocheltree Kennedy.

The boy in the picture is Newark native and Civil War legend, John Clem.

(Photo: Brockway, Images of America – Newark)

> *"Too much cannot be said to the praise of the noble women of Licking County in their patriotic and Christian work for the cause. They forwarded great supplies of sanitary goods to the camps and hospitals in the field, and they soothed the sorrows of the widows and orphans at home."*
>
> **From:** *The History of Licking County O.* by N. N. Hill, Jr., 1881.

During the Revolutionary War and Civil War, it is suspected that hundreds of women disguised themselves as men and joined in the fighting – perhaps even a few from Licking County?

The impact of war can reach far beyond the battlefield. During World War II, millions of women entered the American workforce, many for the first time, to help with the war effort and to keep the economy alive. Women moved into these employment roles as men went overseas to fight. This significantly challenged the traditional and pervasive perceptions of what work women could perform.

Women's involvement in the workforce during WWII set the stage for expanding opportunities for women. In 1950, only 33.9% of women in the United States were employed in the workforce. Today, nearly 60% of women are employed outside the home.

The presence of women in the military has increased substantially as well. In 1950, women made up less than 2% of the military. Today 14%, or more than 198,000 of active duty soldiers are women, 33,500 of whom are officers.

Today there are more than 1.8 million female military veterans. They represent 5% of WWII veterans, 2% of Korean War veterans, 3% of Vietnam veterans and 16% of veterans that served in the

1990-91 Persian Gulf War.

Many Licking County women have made significant contributions to the United States' military efforts. In addition to those who have been inducted into the Ohio Veterans Hall of Fame, there are many other local women with interesting military backgrounds.

Here are just a few:

Mary C. Oliver served in the U.S. Army from 1983 to 1986, with the 101st Airborne Division in Kentucky, and the 2nd Infantry Division in Korea. She is currently the Licking County Veterans' Service Officer. She serves on the Board of Directors of the American Red Cross, and on Congressman Space's Veterans Advisory Council. Mary has testified before a Congressional Subcommittee studying veterans' health care. She is a life member of the DAV Chapter #23, AMVETS Post #345, American Legion Post #85, and the VFW Post #1060. She has served as commander over 31 American Legion Posts in seven counties.

Hebron's Kimberly Carbaugh enlisted in the Air Force in 1982. She completed training in San Antonio, Texas at Lackland Air Force Base. She was later transferred to Grand Forks, North Dakota. Kimberly's assignment was in the field of transportation. She also served on funeral details and in the honor guard.

Ruth Kaiser from Buckeye Lake joined the Navy in 1942. She completed her initial training at a base in Norman, Oklahoma, and then was transferred to work various assignments near Washington, D.C. Ruth worked for an office that prepared orders for sailors. She later worked at Arlington National Cemetery in the casualty office. The casualty office was responsible for notifying families of the death of a related service member.

Former Owens-Corning employee, Lucille Smith joined the Navy in 1942 and became an aircraft mechanic in Tennessee and at Alameda Naval Air Station in California.

Virginia Toffant served as a nurse in the Army during World War II. She was stationed in Europe from 1940-1945, where she enjoyed working in France, Germany, England, and Belgium. It has been reported that she played cards with future president General Dwight Eisenhower in Berlin and had the opportunity to meet the famous General George Patton. Virginia was a graduate of the Mount Carmel school of nursing. She lived her later years in Utica, Ohio.

Pearl Deeds enlisted in the Army Nurse Corps Reserves in 1941 and was sent to the Pacific theater. She served 3 1/2 years in various overseas locations, attaining the rank of Captain. Pearl is now 90 years old and lives in the Licking Valley area.

Township Notes

Harrison Township was home to Leverett Butler who became a noted marksman in the War of 1812. It was said that he could shoot out the flame of a candle from 100 yards. In 1823, he participated in the Great Wolf Hunt to protect livestock from predators. One black bear, one owl, three wolves, forty-nine deer, and 60-70 turkeys were harvested in one day.

Hartford Township was divided on the issue of slavery. There are reports of abolitionists being beaten by mobs for their anti-slavery ideas. Fifty-three men from Hartford Township fought in the Civil War, 20 of them never to return. After the war, 6 Confederate veterans settled in Hartford Township.

Newton Township was home to General John Spencer, who fought as a captain in the War of 1812. Following the war he maintained a militia in Newton Township.

St. Albans Township's Joseph M. Scott enlisted in the Union Army in 1861. As a captain in Company B of the 76th Ohio Volunteer Infantry, he fought in the battles of Shiloh, Corinth, and Milliken's Bend. His residence, known as the Captain Scott House, was moved during the expansion of State Route 161 in 2007-2008 to Johnstown-Alexandria Road on the west side of Alexandria.

From: *A Bicentennial History of Licking County Ohio, 1808-2008, Dan Fleming & Bob Tharp.*

Scrap Drives

During WWII there was a huge need for raw materials in order to meet the demands of manufacturers trying to supply the military with weapons, vehicles and other necessary items. Scrap drives were held across the nation. During these drives, all kinds of items were collected including rubber, newspapers and rags. Metal, of course, was especially valued. Many communities donated everything from their wrought iron fences and streetcar rails to Civil War cannons and statues.

The owner of Baughman Park, located just across the Licking County line near Frazeysburg, donated two brass cannons that were showpieces in the park. In the Kirkersville Cemetery, a large Soldiers Monument with a cannon on top (see below) was torn down and its metal components donated to the war effort.

Industries and War

The extensive involvement of American troops in World War II required citizens and companies alike to focus on the war effort. Citizens were encouraged to express patriotism by purchasing bonds. Individuals and companies were expected to conserve materials for the war effort. Manufacturers across the nation were challenged with the task of retooling and refocusing their efforts to make the products needed by the military. Licking County companies met the challenge. Here are a few examples of contributions to the war effort made by Licking County's industries.

Alcoa

The Aluminum Company of America (Alcoa) factory in Newark participated in the war effort by producing a variety of aluminum products for the military. Aluminum was used extensively in war plane productions.

Charles Schuler Company

The military used zinc plating for a variety of applications during the war, including coating steel cartridges. The Charles Schuler Company of Newark provided zinc plating and fabrication services for the war effort.

Newark Stove

The Newark Stove Company, known as the Wehrle Stove Company prior to 1939, was at one time the largest employer in Licking County. The company's typical product line included more than 65 styles of stoves, a variety of cooking utensils, and fireproof safes. During the war, the company produced land mines, shells, bomb bay doors for aircraft and other products.

Owens-Corning

During World War II, Owens-Corning produced Fiberglas insulation and other products used in the construction of military ships and aircraft.

Pharis Tire & Rubber

Newark's Pharis Tire & Rubber Company, produced tires for military vehicles during the war.

Pure Oil

Beman Gates Dawes founded the Pure Oil Company in Licking County. Pure Oil provided aviation fuel during the war. Beman is also know for establishing Dawes Arboretum in 1929.

E.T. Rugg Company
Known for its lawn mowers and other household equipment, during the war, the company produced land mines, proximity fuses, rope products, and other goods for the military.

Tucker Boiler & Tank Works
Newark's Tucker Boiler & Tank Works manufactured fuel tanks and other products used by the military during the war. Tucker Boiler & Tank Works was purchased by Modern Welding Company, which is still in operation today.

Holophane
Lighting fixtures produced by Holophane were used in military aircraft hangars at bases around the world. Engineers from the company collaborated with government officials to develop special lighting for the factories that would produce atomic bombs. The company also helped develop a clandestine signaling system for the Navy. Equipment was affixed to naval ships that emitted a light that could not be detected by the unaided eye. The signals were used by the Navy to secretly transmit orders to its ships.

The Army and Navy awarded Holophane the Award for Excellence in War Production. The award was issued to only a fraction of the companies involved in the war effort.

Heisey Glass Company
The Heisey Glass Company used large metal molds in its production of glassware. Retired and unused molds were highly valued by the company but were donated to the war effort in both World Wars.

Heisey also manufactured gauges used by the Frankford Arsenal in Philadelphia. The gauges were used primarily for measurement during the production of weapons used in the war. Heisey encouraged the public to support the war by purchasing bonds and was dedicated to ensuring veterans had jobs when they returned.

A.H. Heisey, founder of the Heisey Company, served in the Civil War and fought at such battles as Gettysburg.

Jewitt Car Company
The Jewitt Car Company, which was located in Newark and made interurban railway cars, built the first Red Cross ambulance to be delivered in France after America entered World War I.

In addition to the companies discussed above, Smyth Manufacturing, TAKK Corporation, Western Products, and other Licking County companies provided essential materials for the military during World War II.

Above: WWII Heisey advertisement pledging that employees serving in the military will have their jobs waiting for them when they come home.

Above: Part of a magazine advertisement from Heisey during WWII, featuring glass gauges the company was making for the military.

(Ads courtesy of The National Heisey Glass Museum)

The Draft

During several conflicts, the United States has depended upon conscription (commonly called the "Draft") to recruit personnel for the military. Conscription requires men of eligible age to register for military service and to serve if called. Refusal to serve can result in being sent to jail.

Conscription in the U.S. dates back to 1792 when the U.S. was attempting to build up its military. The first wartime conscription used in the U.S. was during the Civil War, beginning with the Confederacy followed quickly by the Union.

The draft was reinstated during World War I and increased sharply during World War II. Conscription continued through the tumultuous Vietnam era and became divisive in the United States culture. The controversy was fueled by loopholes that allowed members of privileged classes to avoid being drafted into military service by attending college. The draft went to a lottery system beginning on Dec. 1, 1969. The drawing for the last lottery was held on March 12, 1975. Today, the United States' military is composed entirely of volunteers.

Number of men drafted in each war (nationwide):

WWI – 2,810,296
WWII – 10,110,104
Korea – 1,529,539
Vietnam – 1,857,304

Newark Air Force Base

In the 1950s, the Air Force implemented the Heavy Press Program to develop manufacturing processes to more effectively design and build aircraft systems. This program required specialized facilities. A site located at Heath in Licking County was selected to house the Heavy Press Program. Kaiser Aluminum and Chemical Corporation was selected as a partner in the project by the Air Force.

The Heavy Press Program was never actually implemented at Heath but the site was eventually used for inertial guidance system repair and calibration.

The facility brought about an influx of new employees and economic development in the early 1960s. The population of the City of Newark and Village of Heath grew substantially as a result. The station eventually became Heath's largest employer.

In the 1990s, decreasing defense budgets forced base closures across the country. In 1993, the military decided to close the Newark base. Since the base was a significant source of tax revenue for local governments and sales for area businesses, a committee recommended transitioning operations of the base from government control to private industry.

The Heath-Newark-Licking County Port Authority was implemented to oversee the newly privatized base in 1995. Today, the HNLC Port Authority continues to serve as landlord of the former base, now known as the Central Ohio Aerospace and Technology Center. The base is currently home to the Boeing Company, AFMETCAL, Bionetics, the Defense Contract Management Agency, and others. The Boeing Company – Heath, is the largest employer at the base and partners with Atlantic Inertial Systems, Draper, Honeywell, Kearfott, Northrop Grumman, and L3 Communications.

Newark National Guard Armory

The Ohio National Guard has an armory located on Hollar Lane in Newark. It is home to three units, including the 211th Maintenance Company, and about 250 troops.

The armory is also home to the guard's central warehouse for Ohio, which means that all kinds of military equipment passes through the Newark facility. The 211th celebrated its 50th anniversary in 1996.

Gun Manufacturing

At a time when nearly every family owned at least one gun, and military men often used their own rifles for duty, Licking County had at least 30 gunmakers and gunsmiths. Some of the more notable of the gunmakers were Aaron Channel and Joseph Zartman. Newark was also the home of Joseph Rider, famous for his firearm innovations. In 1858, he invented the double-action percussion revolver. He eventually held over 100 patents. Rider was hired by the Remington Company and is known for his Remington-Rider percussion pocket revolver and his rolling block rifle design. By 1890, Rider was considered the wealthiest man in Newark. He owned several pieces of property in town, including the large home on Hudson Avenue where he lived.

Bomber Crash in Newark

On September 8, 1942, an Army B-25 medium bomber crashed into downtown Newark, narrowly missing the County Jail and Courthouse. All six Army men in the plane and two civilians on the ground died in the crash. At least three others on the ground were injured. Two of the Army men on board were soldiers on leave, catching a ride east.

The plane crashed near the intersection of Hudson Ave. and Wyoming Street, just north of where the overpass for St. Rt. 16 crosses Hudson today.

Two of the Army officers on board died when they attempted to parachute out of the falling plane. One plummeted through the roof of an apartment house and the other was found entagled in his partially-opened parachute on the platform of the B&O railroad.

The plane had taken off from Wright Field in Dayton, Ohio, heading for New York. It was carrying no bombs. The cause of the accident was thought to be weather related due to the clouds and rain that day, or due to the structural failure of one wing.

Left: A view of the intersection of Hudson and Wyoming as it looks today (looking north). The B-25 knocked down the far right column of the stone house in the center of the photo. The column was later replaced. There is no marker or any indication that a tragic wartime plane crash took place here.

Veterans Alliance

The Licking County Veterans Alliance performs military funerals for veterans in Licking County at no charge.

The non-profit alliance was founded in 1992 and operates through donations.

The group is made up of veterans from all the local veterans organizations and typically participates in over 150 funerals each year.

Often, the alliance is contacted by the funeral home on behalf of a deceased veteran's family. Generally, fourteen veterans are involved in a funeral and help in whatever way they are needed. Sometimes they serve as pallbearers. As part of the ceremony, five members fire three rifle vollies and then pass out the spent cartridges to family members.

Anyone wishing to request the services of the alliance, or to donate funds to support the group's work, should contact the Licking County Veterans Commission or one of the local veterans groups.

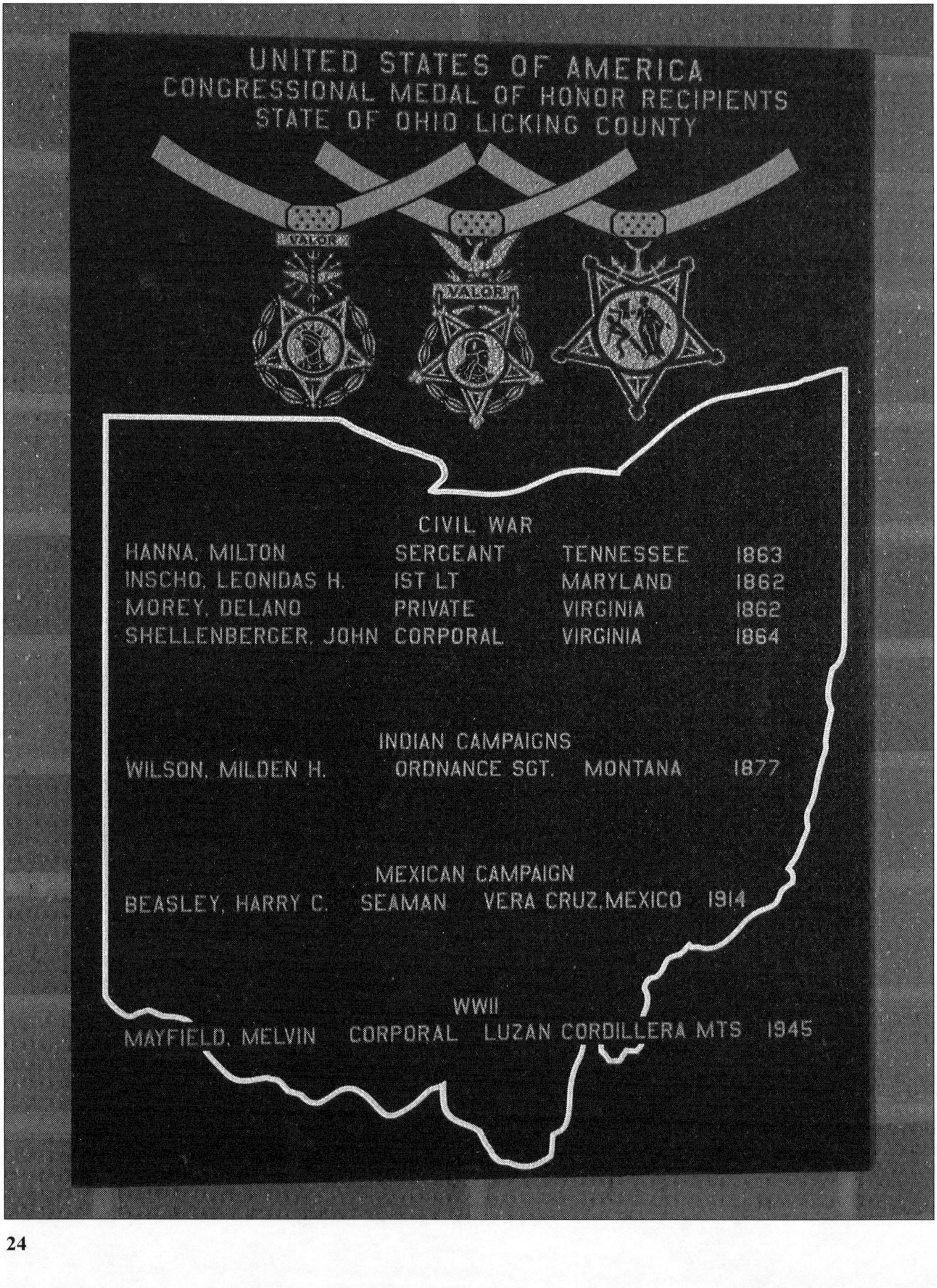
UNITED STATES OF AMERICA
CONGRESSIONAL MEDAL OF HONOR RECIPIENTS
STATE OF OHIO LICKING COUNTY
VALOR
VALOR
CIVIL WAR
HANNA, MILTON SERGEANT TENNESSEE 1863
INSCHO, LEONIDAS H. IST LT MARYLAND 1862
MOREY, DELANO PRIVATE VIRGINIA 1862
SHELLENBERGER, JOHN CORPORAL VIRGINIA 1864
INDIAN CAMPAIGNS
WILSON, MILDEN H. ORDNANCE SGT. MONTANA 1877
MEXICAN CAMPAIGN
BEASLEY, HARRY C. SEAMAN VERA CRUZ, MEXICO 1914
WWII
MAYFIELD, MELVIN CORPORAL LUZAN CORDILLERA MTS 1945

Heroes

Our Medal of Honor Recipients

Seven men who were born, lived, or entered service in Licking County have been awarded the nation's top military decoration: the Medal of Honor.

The Medal of Honor is the highest award for valor in action against an enemy force which can be bestowed upon an individual serving in the Armed Services of the United States. Generally presented to its recipient by the President of the United States of America in the name of Congress, it is often called the Congressional Medal of Honor.

On December 9, 1861, Iowa Senator James W. Grimes introduced S. No. 82 in the United States Senate, a bill designed to "promote the efficiency of the Navy" by authorizing the production and distribution of "medals of honor." On December 21st the bill was passed, authorizing 200 such medals be produced "which shall be bestowed upon such petty officers, seamen, landsmen and marines as shall distinguish themselves by their gallantry in action and other seamanlike qualities during the present war (Civil War)." President Lincoln signed the bill and the (Navy) Medal of Honor was born.

Two months later on February 17, 1862, Massachusetts Senator Henry Wilson introduced a similar bill, this one to authorize "the President to distribute medals to privates in the Army of the United States who shall distinguish themselves in battle." When President Abraham Lincoln signed S.J.R. No. 82 on July 12, 1862, the Army Medal of Honor was born.

There are three different types of Medals of Honor today: the original simple star shape established in 1861 which the Navy, Marine Corps and Coast Guard have retained; a wreath version designed in 1904 for the Army; and an altered wreath version for the Air Force, designed in 1963 and adopted in 1965.

Above: Army Medal of Honor.

Opposite: Plaque honoring Licking County's Medal of Honor winners. The plaque is located at the main entrance to the Licking County Administration Building in Newark, Ohio.

Medal of Honor Statistics
(national totals as of 9/2009)

Total Recipients: 3,448

Double Recipients: 19

Delano Morey

U.S. Civil War, 1862
Rank: Private
Organization: U.S. Army
Company: Company B
Division: 82d Ohio Infantry
Born: Licking County, Ohio
Entered Service At: Hardin County, Ohio
Place & Date: At McDowell, Va., May 8, 1862

Official Citation
After the charge of the command had been repulsed, he rushed forward alone with an empty gun and captured two of the enemy's sharpshooters.

Leonidas H. Inscho

U.S. Civil War, 1862
Rank: Corporal
Organization: U.S. Army
Company: Company E
Division: 12th Ohio Infantry
Born: Chatham, Ohio
Entered Service At: Charleston, W. Va.
Place & Date: South Mountain, Md., Sept. 14, 1862

Official Citation
Alone and unaided and with his left hand disabled, captured a Confederate captain and four men.

Above: Leonidas Inscho's grave marker at Cedar Hill Cemetery in Newark, dedicated in 1988.

Milton Hanna

U.S. Civil War, 1863
Rank: Corporal
Organization: U.S. Army
Company: Company H
Division: 2nd Minnesota Infantry
Born: Licking County, Ohio
Entered Service At: Henderson, Minn.
Place & Date: Nolensville, Tenn., Feb. 15, 1863

Official Citation
Was one of a detachment of 16 men who heroically defended a wagon train against the attack of 125 cavalry, repulsed the attack and saved the train.

John Shellenberger

U.S. Civil War, 1864
Rank: Corporal
Organization: U.S. Army
Company: Company B
Division: 85th Pennsylvania Infantry
Born: (not listed)
Entered Service At: Perryopolis, Pa.
Place & Date: At Deep Run, Va., Aug. 16, 1864

Official Citation: *Captured the flag.*

Milden H. Wilson

Indian Wars, 1877
Rank: Sergeant
Organization: U.S. Army
Company: Company I
Division: 7th U.S. Infantry
Born: Huron County, Ohio
Entered Service At: Newark, Ohio
Place & Date: At Big Hole, Mont., Aug. 9, 1877

Official Citation
Gallantry in forming company from line of skirmishers and deploying again under a galling fire, and in carrying dispatches at the imminent risk of his life.

Harry C. Beasley

MEXICAN CAMPAIGN, 1914
Rank: Seaman
Organization: U.S. Navy
Company: (not listed)
Division: (not listed)
Born: Newark, Ohio
Entered Service At: (not listed)
Place & Date: (place not listed) Aug. 15, 1914

Official Citation

On board the USS Florida for extraordinary heroism in the line of his profession during the seizure of Vera Cruz, Mexico, April 21, 1914.

> CONFLICTING INFORMATION - CIVIL WAR
> Some sources list a Medal of Honor reciepient named Daniel I. Holcomb as being born in "Hartford" in Licking County, Ohio. However, it seems more likely that Holcomb, who served with the 41st Ohio Volunteer Infantry, which was organized in Cleveland, was from another area called "Hartford" in Trumbull County, Ohio.

Melvin Mayfield

WORLD WAR II, 1946
Rank: Corporal
Organization: U.S. Army
Company: Company D
Division: 20th Infantry, 6th Infantry Division
Born: Salem, W. Va.
Entered Service At: Nashport, Ohio
Place & Date: Cordillera Mountains, Luzon, Philippine Islands, Aug. 31, 1946

Official Citation

He displayed conspicuous gallantry and intrepidity above and beyond the call of duty while fighting in the Cordillera Mountains of Luzon, Philippine Islands. When two Filipino companies were pinned down under a torrent of enemy fire that converged on them from a circular ridge commanding their position, Cpl. Mayfield, in a gallant single-handed effort to aid them, rushed from shell hole to shell hole until he reached four enemy caves atop the barren fire-swept hill. With grenades and his carbine, he assaulted each of the caves while enemy fire pounded about him. However, before he annihilated the last hostile redoubt, a machine gun bullet destroyed his weapon and slashed his left hand. Disregarding his wound, he secured more grenades and dauntlessly charged again into the face of point-blank fire to help destroy a hostile observation post. By his gallant determination and heroic leadership, Cpl. Mayfield inspired the men to eliminate all remaining pockets of resistance in the area and to press the advance against the enemy.

Left: After Harry Beasley's military service, he joined the Newark Ohio Police Department.

On June 30, 1931, Patrolman Beasley confronted several men trying to steal a safe from a shoe store in downtown Newark. They exchanged fire and Beasley was shot multiple times. He died from his wounds a few days later.

A memorial to Beasley, including his Medal of Honor and other service medals are on display in the lobby of the Newark Police Station.

Ohio Veterans Hall of Fame

Licking County Members

The Ohio Veterans Hall of Fame seeks to recognize veterans for their countless contributions to society after honorable military service. Inductees are honored during a Hall of Fame ceremony every year in November.

The following is a list of the Licking County veterans who have been inducted into the Hall of Fame, listed in order of their induction.

(Although not listed here, all Medal of Honor recipients from Ohio were inducted in 1992).

William B. Woods

United State Army - Civil War

Inducted into the Hall of Fame in 1994.

The 46th Justice of the U.S. Supreme Court. Elected to Ohio House of Representatives and became Speaker. Served at Shiloh and Vicksburg and with General Sherman. Practiced law in the South. President Grant appointed him to the Fifth Circuit Court from Georgia. President Hayes nominated him to the Supreme Court of the United States. Confirmed by the U.S. Senate. *A photograph of Woods can be seen on page 12 of this book.*

Esther E. Fraizer

United States Navy - World War II

Inducted into the Hall of Fame in 1996.

Held many posts in the American Legion at both Post and District levels. Worked on Buckeye Boys State and visited veterans in nursing homes. Volunteered with Special Olympics at Denison University. Was active in American Association of Retired People (AARP) and the Ohio Child Conservation League. Volunteered with the Salvation Army Soup Kitchen. As "Grandma Esther," read to elementary school children for several years.

Nora M. Kendall Noble

United States Air Force - Vietnam

Inducted into the Hall of Fame in 1996.

First nurse to make transition from nurse corps to the line when she became base commander at Newark Air Force Base. Following military career, established a school in Newark without funds or material - the school was accredited 19 months later (unheard of in the State of Ohio). Involved with community organizations including: Licking County Aging Program, Service Corps of Retired Executives (SCORE), Shiloh Missionary Baptist Church, Private Industry Council (PIC), Young Women's Christian Association (YWCA), United Way, Girl Scouts of America and Board of Visually Impaired. Vice President, Ohio State Medical Board.

John L. Clem

United State Army - Civil War

Inducted into the Hall of Fame in 1997.

Ran away from home and tried to join the Union Army when he was not yet 10 years old in the spring of 1861. Tagged along with the 22d Michigan Regiment that adopted him as their mascot. Officially enlisted in the 22d Michigan Regiment in May 1863. Known as the "Drummer Boy of Chickamauga" and as "Johnny Shiloh". President Grant commissioned him a 2d Lieutenant in 1871. Retired from the Army in 1915 with the rank of Brigadier General; was the last Civil War veteran on the Army rolls at the time of his retirement.

Mary O. Glover

U.S. Women's Army Corps - World War II

Inducted into the Hall of Fame in 2001.

Member and past chaplain of the Women's Army Corps Veterans' Association Chapter 3; Life member of American Legion Post 85 and the AMVETS Post 345; Completed medical and surgical technician training and provided medical care to returning POWs and wounded soldiers; PTA president for Kettering Intermediate School; Volunteer for the Newark City Schools Mentoring Program; Volunteer with the Salvation Army Soup Kitchen; Accomplished public speaker mainly talking on a woman's role in the United States Army; Provides humor for ill children and adults as "Poppy" the clown.

Edith Doctor Jobe

United States Army

Inducted into the Hall of Fame in 2003.

Women's Army Corps; Women's Army Auxiliary Corps; Volunteer Girl Scouts of America - Licking County; American Legion Post Commander 1998-2002; American Legion Post Chaplain and delegate - Licking County; President of American Legion Auxiliary; Tutor for Mokena Public Schools - Richland County, Illinois; First Native American female school bus driver in Licking County; Honor Guard for American Legion Post # 398.

Donald I. Jakeway

United States Army - World War II

Inducted into the Hall of Fame in 2009.

While serving in World War II, pledged to serve the elderly, children and others in need if he returned from the war. Supports athletics and charities, served on the village council and established a high school scholarship program. Past Commander of American Legion Post #254, Veteran of the Year in 2003, Legionnaire of the Year in 2004, and received Licking County's Inscho-Beasley-Mayfield Award in 2008.

Above: The inscription on this monument in Kirkersville Cemetery reads: "B.F. Wilcox, Co. B, 135th Reg., O.N.G., Wounded at North Mountain VA. Died in hospital at Claysville, MD., Aug. 4, 1864, Aged 21 yrs., 6 M., 9 D."

Inscho-Beasley-Mayfield Award

For Licking County Veteran of the Year

The **Inscho-Beasley-Mayfield Award** is presented by the Licking County Veterans Service Commission each year for outstanding community service. It is named after the three Medal of Honor recipients who were born, lived and died in Licking County. It was established by former president of the commission, Glenn Cunningham. Each veterans organization in Licking County is invited to nominate one of their members.

The following criteria apply:

The nominee must be a living veteran who has not received the award previously. He or she must reside in Licking County and be a member of the nominating organization. A copy of the veteran's DD Form 214 must accompany the nomination.

The commission sets a deadline to turn in nominations and then the five-member commission reviews them and votes privately. A dinner is held to recognize all the nominees. Officers and their spouses, elected officials, and the commission and staff all attend. Each nomination is read so that the audience can share in the outstanding work that these volunteers do for our community. Each nominee receives a small eagle trophy. The president then announces the recipient of the Veteran of the Year Trophy.

Inscho-Beasley-Mayfield Award

Award Recipients:

Reverend Robert Knox - 1999
Nominated by American Legion Post #85

Daniel L. Dupps - 2000
Nominated by American Legion Post #771

John C. Jacobs - 2001
Nominated by American Legion Post #85

Philip Gregg - 2002
Nominated by American Legion Post #92

Frank Wright - 2003
Nominated by VFW Post #1060

Joseph E. Schweitzer - 2004
Nominated by the Marine Corps League

James "Jim" Welsh - 2005
Nominated by American Legion Post #398

Vince Becker, Jr. - 2006
Nominated by AMVETS Post #345

William "Bill" Balser - 2007
Nominated by American Legion Post #804

Donald I. Jakeway - 2008
Nominated by American Legion Post #254

Left: Grave of Confederate soldier, Corporal John J. Wilson in North Lawn Cemetery, Utica. Curiously, the grave is marked with both a Confederate marker and a G.A.R. (Grand Army of the Republic) marker.

Wilson was a member of Co. F, 23rd Virginia Infantry. Local lore says that Wilson was a prisoner on a train passing through Utica on his way to Johnson's Island prison camp at Lake Erie. He supposedly died on the train and was buried in this cemetery, which is located across the road from the railroad tracks.

"The Virginia Regimental Series" by Thomas M. Rankin offers the following information about Corporal Wilson: "Corp., Co. F b. Goochland Co. Enl 5/20/61 at Manakin for 1 yr. 4th corp., on sick leave 7/61 until detailed acting orderly sgt. 5/2/62. Discharged for disability 7/23/62. Age 36, millwright, 6', dark complex., dark hair, dark eyes. Buried City Cemetery, Utica, Ohio."

Corporal Wilson apparently served a little over a year and was sick much of that time. He was then discharged for disability. Nothing indicates that he was involved in a battle or captured. How he ended up in Utica, Ohio, is still somewhat of a mystery.

Left: A veteran who served in WWII, Korea and Vietnam. This marker, in Eden Church Cemetery, marks the resting place of Charles H. Braid.

Sanford Cemetery

Other Notable Licking County Veterans

Serving in the military is often a life-changing experience. Every veteran has a story regardless of whether he or she was a private or general, or served in peacetime or war. There are thousands of stories that could be told about the lives of veterans from Licking County. Below are some brief notes about just a few of the more well-documented veterans.

Major-General Charles Griffin

Charles Griffin was born in Granville, Ohio in 1825 and attended Kenyon College in Gambier. He served in the Mexican-American War and fought the Navajo Indians in New Mexico Territory. Griffin later served as a Union general during the Civil War, participating in many campaigns and was present during Robert E. Lee's surrender at Appomattox Courthouse.

General Griffin Road, west of Granville is named after this notable Licking County veteran.

Above: Gen. Charles Griffin.

Captain Joseph M. Scott

Joseph Scott was born in Utica, Ohio in 1830 and served in the Civil War. He fought in the Battle of Shiloh, the Battle of Milliken's Bend and others. He was a vocal abolitionist and espoused the cause of the Prohibition Party. He ran for several political offices including sheriff and lieutenant governor. Scott died in 1919.

The Victorian Italianate style house he built near Alexandria, Ohio has recently been moved to save it from destruction due to the widening of Routes 161 & 37.

Above: Cpt. Joseph M. Scott.
(Photo: Brister, Centennial History, 1909)

Major N. Bostwick

Bostwick was a Licking County farmer who enlisted in the Union Army during the Civil War. He was captured at the Battle of Atlanta and shot in the face. While in southern prisons he suffered inhumane treatment and near starvation. He and several others eventually made a harrowing escape through the mountains.

The Woods Brothers

Newark brothers, Charles Robert and William Burnham Woods, both served in the Civil War. Charles was a colonel in the 76th Ohio Volunteer Infantry. He retired in 1874 as a major-general.

William served as Newark's mayor before joining his brother in the 76th. Later, he was appointed to the U.S. Supreme Court by President Hayes.

Opposite: One of dozens of small cemeteries scattered across the fields and woods of Licking County with headstones that are broken, unreadable, overgrown, piled in corners or stacked against trees. No doubt, many of these stones once marked the graves of veterans – now forever lost.

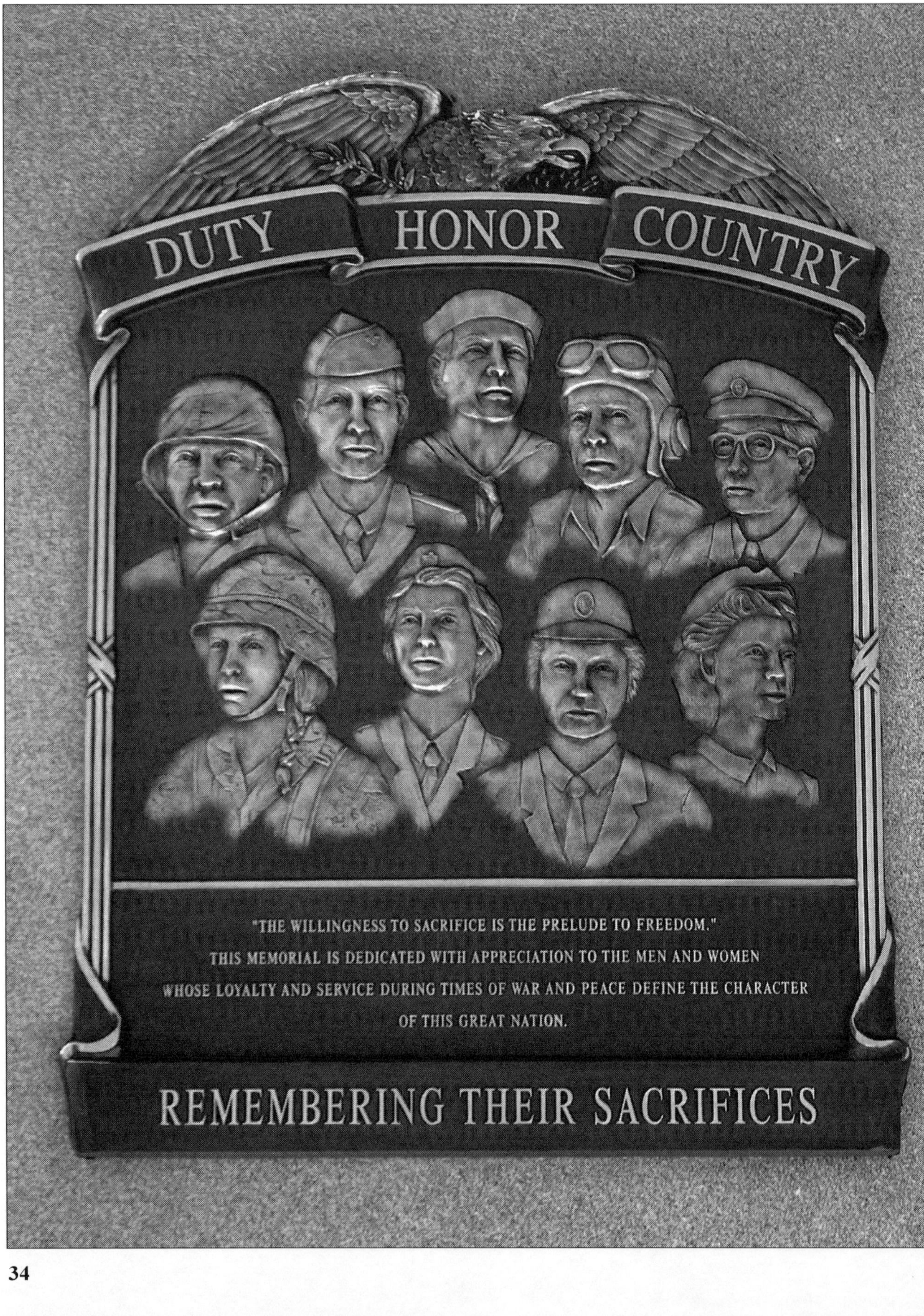
DUTY HONOR COUNTRY
"THE WILLINGNESS TO SACRIFICE IS THE PRELUDE TO FREEDOM."
THIS MEMORIAL IS DEDICATED WITH APPRECIATION TO THE MEN AND WOMEN
WHOSE LOYALTY AND SERVICE DURING TIMES OF WAR AND PEACE DEFINE THE CHARACTER
OF THIS GREAT NATION.
REMEMBERING THEIR SACRIFICES

MEMORIALS

Veterans' Monuments, Statues & Plaques in Licking County

Some memorials to Licking County veterans are in prominent places and are relatively well known. Others are located in spots less often visited, such as old cemeteries, churches, government buildings and small town parks.

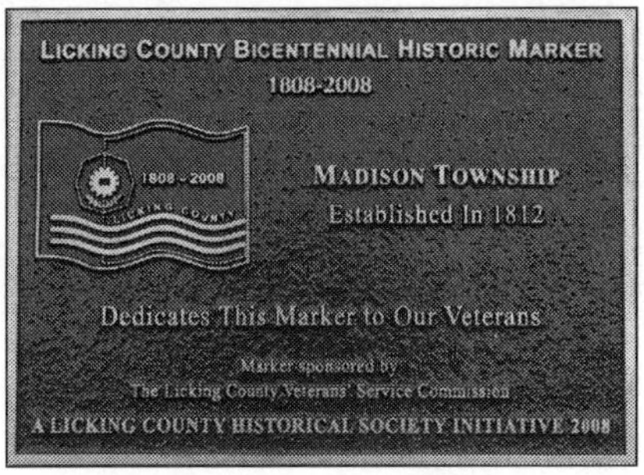

Above: One of the Licking County Bicentennial Historic Markers made for each township in the county during the county's bicentennial in 2008. The plaques, which are dedicated to veterans, commemorate the year each township was established. The marker project was a Licking County Historical Society initiative, sponsored by The Licking County Veterans' Service Commission.

Left: Statue of Civil War soldier now located in front of the V.F.W. on Forry Avenue in Newark, Ohio.

Opposite: Memorial plaque on the veterans monument in Cedar Hill Cemetery, Newark.

As might be expected, major wars tend to have more numerous memorials. Some of the larger cemeteries have special sections devoted to veterans of specific wars. But not all veterans monuments, statues, plaques, etc. in Licking County are widely known or documented. Hopefully this visual guide will help bring some of those more obscure markers to light.

Above: Plaque dedicating the Veterans Walk of Honor, 1996.

Right: Plaque dedicating Veteran's Park in 1979, mounted on large piece of flint from Flint Ridge.

Below: Monument honoring the unknown soldier. Dedicated in 1940 by the V.F.W.

Above: Detail of Civil War monument at north end of Veterans Walk of Honor.

Left: Veterans Walk of Honor, view looking north.

Below: Civil War marker dedicated in 1933 to the memory of Ayers B. Adams.

Left: Stone monument placed by the Daughters of Union Veterans of the Civil War.

Sixth St. Park, *continued*

Above: Lone gravestone in Sixth St. Park marks the grave of Wm. Burgess, who died in 1837. The obelisk is located on an earthen mound, which local lore says may conceal a pile of headstones placed there when the old cemetery was leveled during the late 1800s. Why this one headstone was left standing is not known.

Right: Markers listing the names of veterans of the Revolutionary War and War of 1812 known to have been buried in Sixth Street Cemetery.

Opposite Top: Brick sculpture in Sixth Street Park featuring scenes from all wars.

Sixth Street Park is a Graveyard

Sixth Street Park, located on Sixth Street between West Church and West Main Streets, was originally Newark's main cemetery. During the early years of Newark, this cemetery, which at the time was on the outskirts of town, held many graves. At least six Revolutionary War veterans and five veterans of the War of 1812 are known to have been buried there.

Eventually, the cemetery became filled to capacity and burials were no longer allowed. Then, in 1875, following years of neglect, the city ordered all remains moved to Cedar Hill Cemetery, giving citizens only about three months to comply. This short notice, along with the fact that many of the deceased no longer had relatives in the area, led to an almost complete lack of compliance.

At some point it appears that the city removed the headstones but did not move most of the graves. Some believe the headstones may have been stacked and covered with dirt, creating the mound now located along the east edge of the cemetery.

According to writings by the late local historian, W. T. Huff, only a small number of graves were actually moved. Huff estimates that 300 or more bodies are still there. The late local historian Bud Abbott estimated that there "... are 282 known and about a hundred unknown persons buried there."

Over the years, bones and headstones have been exposed during various construction projects in what we now call Sixth Street Park. Graves were disturbed in the park in 1879 when West Church Street was extended west of Sixth Street. A skeleton was dug up during the construction of a fountain in the park in 1909. In 1973 and again in 1992, parts of skeletons and/or headstones were found during work on the National Heisey Glass Museum facility. In 1996, the graves of a man and his son were uncovered during the construction of steps in the park.

There have been several court cases related to the ownership and use or misuse of this property. The Ohio Supreme Court ruled in 1915 that the property "... shall be preserved and improved as a graveyard, but not otherwise...." Obviously, that has not happened.

Today, Sixth Street Park is a public park featuring historic buildings that were moved there and restored. The park is also home to the National Heisey Glass Museum, the office and collections of The Licking County Historical Society and the Veterans Walk of Honor.

Sadly, just beneath the asphalt and grass of this attractive park lies the remains of perhaps hundreds of Newark's earliest citizens, including some of the very veterans the park now honors.

Below: A news item from the Advocate, October 29, 1875 tells about the removal of one of the graves at Sixth Street.

ON last Monday, the remains of John Taylor, after an interment of 43 years, were exhumed from their resting place in the old grave yard. At the bottom, and about the middle of the grave, an ounce and a half of metallic mercury was found. The following circumstance will account for it being there: On the night of his death, August 9, 1832, Dr. Brice, the physician in attendance, admininistered quicksilver with the view of its acting mechanically in removing an obstruction (*invagination,*) of the bowels, which must if not relieved end fatally. The remedy, however, did not have the desired effect, as he soon after expired. The element remained in this situation for 43 years without the slightest deterioration, while those of the man, the lime of the bones excepted, returned to their respective places in nature. J. W. A.

Right Top: Marker honoring the Grand Army of the Republic dedicated by the National Woman's Relief Corps, Dept. of Ohio, 1933.

Right Middle: A plaque honoring Clara Barton, founder of The American Red Cross, also by the Woman's Relief Corps.

Right Bottom: This plaque originally marked an oak tree taken from the Chickamauga battlefield and planted in Sixth St. Park in 1938 by the Lemert Relief Corps. in honor of John Clem. Lemert Post 71 of the G.A.R. was located in Newark.

Below: Plaques from 1937 dedicated to Col. Charles Young and Hon. Wm. A. Ashbrook, and a plaque in honor of Frank D. French, the county's last Civil War veteran. French died in 1939 at age 95. The plaques are located alongside the brick walkway.

Heath – Moundbuilders Park (Great Circle Earthwork)

Left: Two of more than 20 similar stone markers embedded in the brick walkway at Sixth Street Park.

Newark – Trinity Church

Above: A plaque in Trinity Episcopal Church honoring church members who served in World War II.

Below: A plaque in Trinity Episcopal Church honoring church member, William Burnham Woods, a Civil War general who later became a Supreme Court Justice.

Left Middle & Bottom: Two views of a 1947 plaque honoring WWII veterans, located at the Great Circle Earthwork near the Route 79 entrance.

DUTY HONOR COUNTRY
REMEMBERING THEIR SACRIFICES

IN MEMORY OF
ALL AMERICAN VETERANS
This memorial honors all American veterans who, although separated by generations, shared a common, undeniable goal - - to valiantly protect our country's freedoms.
The memories of these American veterans will continue to live on whenever and wherever democracy exists.
The American veteran - - forever a symbol of heroism, sacrifice, loyalty and freedom.

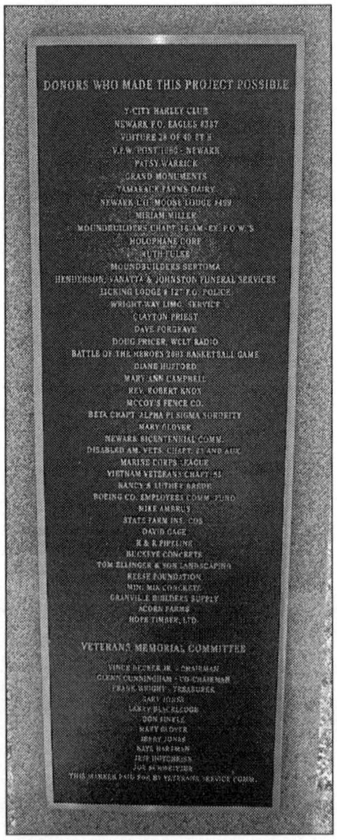
DONORS WHO MADE THIS PROJECT POSSIBLE
NEWARK F.O. EAGLES #387
PATSY WARRICK
GRAND MONUMENTS
TAMARACK FARMS DAIRY
MIRIAM MILLER
HOLOPHANE CORP
MOUNDBUILDERS SERTOMA
HENDERSON, VANATTA & JOHNSTON FUNERAL SERVICES
WRIGHT-WAY LIMO. SERVICE
CLAYTON PRIEST
DAVE FORGRAVE
DOUG PRICER, WCLT RADIO
DIANE HUFFORD
MARY ANN CAMPBELL
REV. ROBERT KNOX
MCCOY'S FENCE CO.
MARY GLOVER
NEWARK BICENTENNIAL COMM.
STATE FARM INS. COS
DAVID GAGE
BUCKEYE CONCRETE
REESE FOUNDATION
GRANVILLE BUILDERS SUPPLY
ACORN FARMS
HOPE TIMBER, LTD.
VETERANS MEMORIAL COMMITTEE

Left: Monument with sundial on top, honoring Revolutionary War veterans.

Bottom: Civil War-era cannon in a section of the cemetery dedicated to Civil War soldiers.

Opposite: Views of the Veterans Monument in Cedar Hill.

Newark – Memorial Gardens Cemetery

Above: A plaque in the Newark Memorial Gardens cemetery along Rt. 13 north of Newark, dedicated in 1956 to veterans of all wars.

Newark – Hollar Cemetery, St. Rt. 13 North

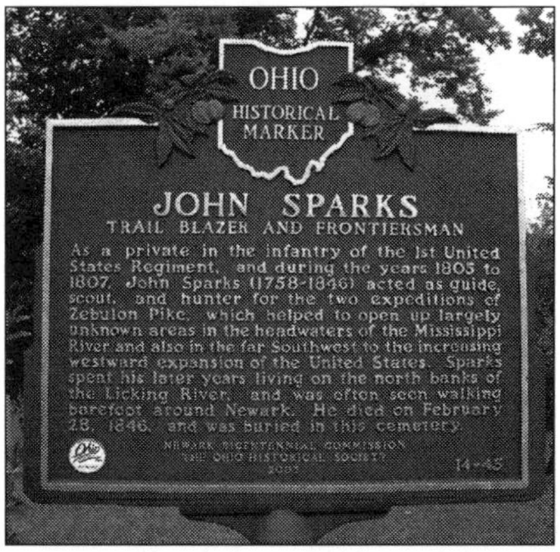

Above: Hollar Cemetery marker honors veteran, John Sparks.

This Page: Statues and plaques in front of the V.F.W. on Forry Avenue in Newark.

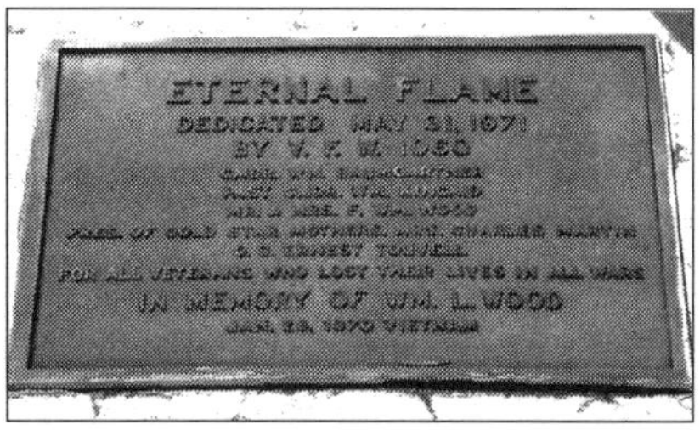

The soldier on the left and sailor on the right were originally mounted on Newark's Auditorium Theater. The WWI soldier in the center (and on the front cover of this book) was created by Ernest Moore Viquesney, a veteran of the Spanish-American War.

Below: "Spirit of the American Doughboy," is the middle statue's official name. The statue was originally made for the town of Nashville, Georgia. It was given to Newark around 1928 by the Newark post of the V.F.W. and placed in Sixth Street Park where it stood for many years (as shown below).

Newark – Vietnam Veterans Memorial Treescape

Left & Above: Vietnam Veterans Memorial Treescape, along St. Rt. 79, can be viewed by heading north on 79 from W. Main St., then stay in far left lane. Treescape is located on the left immediately after the W. Church Street overpass.

Newark – American Legion Post #85, S. Sixth Street

Left & Below: Plaques honoring Devon Seymour and Joseph Lee Nice, local U.S. Marines killed in Iraq. The plaques are located at the southwest corner of the American Legion parking lot. Photo below shows a section of veterans' bricks at the front of the parking lot in the "Reverend Bob & Dorothy Knox Memorial Flag Display," dedicated in 2001.

Heath – Geller Municipal Park

Right Top & Bottom: Views of veteran's monument atop a hill in Geller Park in Heath.

Heath – City Pool

Below: Memorial markers at the base of trees along the west edge of the Heath City Swimming Pool parking lot.

Heath – Road Sign

Below: Heath street named for two of the veterans above. Heath's Sgt. Simpson Parkway honors Sgt. Douglas Simpson.

Left: Monument in Hebron Cemetery.

Below: Views of the monument at Evans Park in Hebron.

Above: Marker near entrance of old cemetery along Rt. 62 near downtown Johnstown.

Below: Plaque in old cemetery honoring eight Revolutionary War soldiers buried there.

Right Top: Star-shaped monument with flags and cannon on west side of Public Square in Johnstown.

Right Bottom: Detail of the plaque at base of the monument on the Public Square.

PATASKALA – COMMUNITY PARK AND POLICE STATION

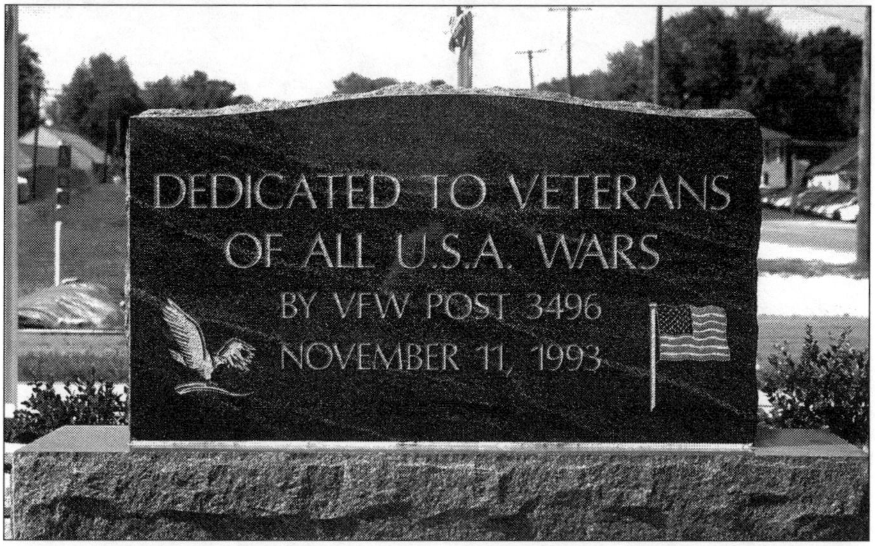

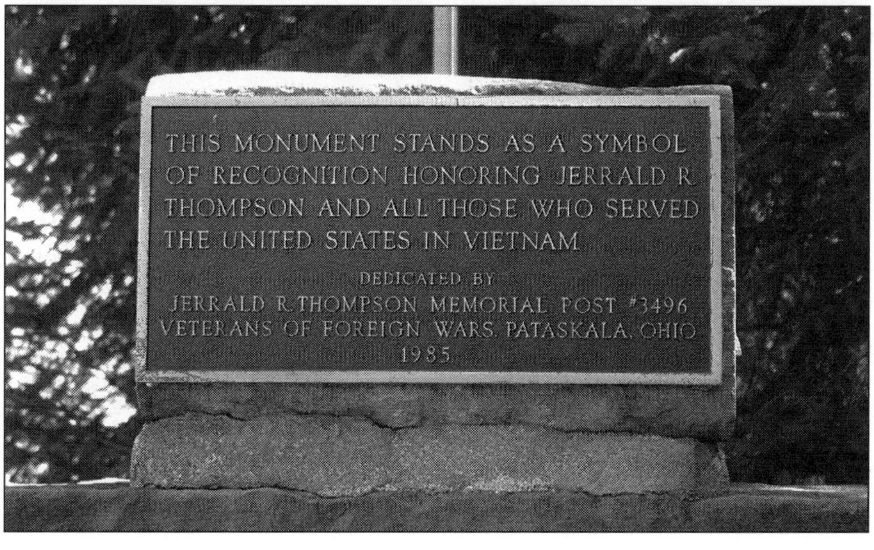

Left Top: Memorial near railroad crossing along Route 16 in Pataskala.

Left Middle: Detail of the monument at the center of the Pataskala memorial.

Left Bottom: Plaque in front of the Pataskala Police Station in recognition of Jerrald R. Thompson and all Vietnam veterans.

CROTON – TOWN SQUARE

Above: Monument in Croton honoring the Croton-area men killed in the Civil War.

Right Top: Large plaque inside Licking Valley High School.

Below & Right Bottom: Plaque mounted on a rock near the entrance of Smith Chapel Cemetery on Reform Road near Hanover.

Above: A plaque at the gate of the Alexandria Cemetery.

Left Top, Bottom & Below: Views of the memorial to veterans of all wars, located in the Alexandria Cemetery.

Right Top & Bottom: Views of monument near entrance of Fairmount Cemetery, east of Jacksontown, north of Rt. 40 on Fairmount Road (Twp. Road 323).

Note the prehistoric burial mound at left edge of photo.

DAWES ARBORETUM

Above: Cemetery marker on the grounds of Dawes Arboretum, St. Rt. 13 between Newark and Jacksontown.

Above & Left: Views of a plaque and memorial trees at Infirmary Mound Park along St. Rt. 37 south of Granville.

HOMER – CEMETERY

Below: Veterans marker in the cemetery at Homer.

Right Top & Middle: Veterans' plaque at Perryton Cemetery.

Below & Right Bottom: Headstone and marker for Pvt. William Lugenbeal, a Licking County man who survived the Civil War, including a stay in the infamous Andersonville Prison, only to face another horror on his way home from the war. He was among the 2,300 ex-prisoners of war on the paddlewheeler Sultana when it exploded on the Mississippi River, killing about 1,800 of the men on board. According to legend, Lugenbeal survived by killing the boat's alligator mascot with his bayonet and using the gator's wooden crate to float on. Lugenbeal lived to the age of 82, dying in 1916.

Left & Below: A plaque mounted on a large boulder at the southeast corner of the intersection of Main Street (St. Rt. 13) and Spring Street in Utica.

Left: Plaques honoring veterans of Korea, Vietnam, Desert Storm and the Persian Gulf, located beide the American Legion building at Spring and Washington Streets.

DRUMMER BOY

John Lincoln Clem

Of all the notable Licking County men and women who have served in the military, there is one whose name is truly legendary: Johnny Clem.

Also known as "The Drummer Boy of Chickamauga," and "The Drummer Boy of Shiloh," Clem's story has become part of our national narrative. Part fact, part folklore, Johnny Clem is without a doubt the best known drummer boy of the Civil War.

John Joseph Klem (he later changed his name to John Lincoln Clem) was born August 13, 1851 and was christened at St. Francis DeSales Church in Newark, Ohio.

Clem ran off to join the Union Army at age 9 but was initially turned down. He tagged along anyway and was finally taken in as a drummer boy.

When it comes to Clem's war adventures it is difficult to separate fact from fiction. It is said that his drum was smashed by a Confederate cannonball while he was playing it at the Battle of Shiloh. His experiences there were portrayed in the 1963 Disney movie "Johnny Shiloh." Today there is some debate among historians as to whether Clem was even at the Battle of Shiloh.*

At the Battle of Chickamauga, Clem reportedly shot a Confederate colonel who was trying to capture him. After the battle he was promoted to sergeant, becoming the youngest non-commissioned officer in Army history. He was discharged in 1864 and re-joined the Army after graduating from high school in 1870.

In 1871, President Grant appointed Clem to the rank of 2nd lieutenant. He remained in the Army until 1915 when he retired as a major-general and the last Civil War veteran still on active duty in the armed forces. General Clem died in Texas on May 13, 1937 and is buried in Arlington National Cemetery.

Above: John Clem as a young man. In later years he became a strong advocate for veterans.

Opposite: John Clem at age 12, while serving as a sergeant in Co. C, 22d Michigan, 1863.

***Editor's note:** There are many discrepancies in print and online regarding the dates and details of John Clem's life and military service.

Note: These two pages offer a fascinating look into the life of John Clem as told by his sister, Lizzie. They are from Volume II of Henry Howe's "Historical Collections of Ohio," published in 1896.

The Drummer Boy of Shiloh.—Newark takes pride in her reputation of having supplied the youngest and smallest recruit to the Union army, and in the person of JOHNNIE CLEM, sometimes called the Drummer Boy of Shiloh, and sometimes of Chickamauga. Lossing says he was probably the youngest person who ever bore arms in battle. His full name is John Winton Clem, but the family spell the name Klem and not Clem. He was born in Newark, August 13, 1851, and ran away from home when less than ten years of age and enlisted as a drummer boy in the army; was in many battles and won singular distinction.

Johnnie Clem's parents were French-Germans, his mother from Alsace. His father was a market-gardener and huckster, and used to send Johnnie, accompanied by his sister, Lizzie (now Mrs. Adams), two years younger, from house to house to sell vegetables. Johnnie was a universal favorite with the people, being a bright, sprightly boy, and very small of his age—only thirty inches high.

The family are now living in garden-like surroundings on the outskirts, on the Granville road, where I went to have an interview to get the facts of his history. I knocked at the side-door of an humble home. A sturdy, erect, compact little woman answered my knock, and to my query replied, "I am his sister and can tell you everything. Please take a seat and I'll be ready in a few moments." She was the Lizzie spoken of above. It was the kitchen I was in: two young children were by her side, and some pies, with their jackets on, on the table about ready for the oven, and only requiring the trimming off of the overhanging dough, which she did dextrously, twirling them on the tips of her uplifted fingers during the operation. Placing them in the oven, and then "tidying up things a little," she took a seat and thus opened up her story for my benefit, while the children in silence looked at me with wondering eyes and listened also:

LIZZIE'S NARRATIVE.

It being Sunday, May 24, 1861, and the great rebellion in progress, Johnnie said at dinner-table: "Father, I'd like mighty well to be a drummer boy. Can't I go into the Union army?" "Tut, what nonsense, boy!" replied father, "you are not ten years old." Yet when he had disappeared it is strange we had no thoughts that he had gone into the service.

When dinner was over Johnnie took charge of us, I being seven years old and our brother, Lewis, five years, and we started for the Francis de Sales Sunday-school. As it was early he left us at the church door, saying, "I will go and take a swim and be back in time." He was a fine swimmer. That was the last we saw of him for two years.

The distress of our father and step-mother at Johnnie's disappearance was beyond measure. Our own mother had met with a shocking death the year before: had been run over by a yard engine as she was crossing the track to avoid another train. No own mother could be more kind to us than was our step-mother. Father, thinking Johnnie must have been drowned, had the water drawn from the head of the canal. Mother travelled hither and yon to find him. It was all in vain. Several weeks elapsed when we heard of him as having been in Mount Vernon; and then for two years nothing more was heard and we mourned him as dead, not even dreaming that he could be in the army, he was so very small, nothing but a child.

It seems he went up on the train to Mount Vernon and appeared next day at the house of Mrs. Dennis Cochrane, an old neighbor of ours. He told her that his father had sent him there to peddle vegetables which were to come up from Newark. None arriving, Mrs. Cochrane surmised the truth, and at the end of the week, fearful he would escape, fastened to him a dog chain and put him in charge of a Newark railroad conductor to deliver to his home, which he could readily do as it was near the depot. On his arrival here he worked on the sympathies of the conductor to let him go free, saying his father would whip him dreadfully if he was delivered to him. This father wouldn't have done—he would have been but too glad to have got him.

The train carried him to Columbus, where he enlisted as a drummer boy in the 24th Ohio. Finding an uncle in that organization he left it and went as a drummer boy in the 22d Michigan. He was an expert drummer, and being a bright, cheery child, soon made his way into the affections of officers and soldiers.

He was in many battles: at Shiloh, Perryville, Murfreesboro', Chattanooga, Chickamauga, Nashville, Kenesaw, and others, in which the army of the Cumberland was engaged. He was at one time taken prisoner down in Georgia. The rebels stripped him of everything, his clothes, his shoes, his little gun—an ordinary musket, I suppose, cut short—and his little cap. He said he did not care about anything but his cap. He did want to save that, and it hurt him sorely to part with it, for it had three bullet holes through it.

When he was exchanged as a prisoner he came home for a week. He was wasted to a skeleton. He had been starved almost to death. I was but a little thing then, but I never shall forget his dreadful corpse-like aspect when the carriage which brought him stopped at our door. He seemed like as if he was done up in a mass of rags. There were no soldier clothes small enough to fit him, and he was so small and wan and not much larger than a babe, about thirty inches high, and couldn't have weighed over sixty pounds.

He returned to the army and served on the staff of General Thomas until the close of the war. After it, he studied at West Point, but could not regularly enter as a cadet on account of his diminutive size. General Grant, however, commissioned him as a Lieutenant. He is now (1886) Captain of the 24th U. S. Infantry, and is stationed at Fort McHenry, Md. He is still small: height, only five feet, and weight, 105 pounds. He married, May 24, 1875, Annita, daughter of the late General Wm. H. French, U. S. A. Like her husband, she is under size, short and delicate; can't weigh over seventy pounds. They have had six children, only one of whom is living.

I have told you of the dreadful death of our mother, run over by a yard engine. My brother Louis, five years old on that noted Sunday, also came to a shocking end. I think father will never get over mourning for him. He grew to be very tall, full six feet, but of slender frame and feeble health. He was off West on a furlough for his health when he went with Custer, as a guest, on his last ill-fated expedition, and was with the others massacred by the Sioux, under Sitting Bull, in the battle of Little Big Horn, in Montana, June 25, 1876.

On closing her narrative Mrs. Adams showed me a portrait of her brother as a captain. He is a perfect blonde with large blue eyes, large straight nose, and a calm, amiable expression. Another as a child standing by the side of General McClennan, who looks pleased, the natural result of having such a sweet-looking little fellow by him. He was a great favorite with all the generals, as Grant, Rosecrans and Thomas, the latter keeping up with him a fatherly correspondence as long as he lived.

To the foregoing narrative from Mrs. Adams we have some items to add of his war experiences, from an equally authentic source.

When he joined the 22d Michigan, being too young to be mustered in, he went with the regiment as a volunteer, until at length he was beating the long roll in front of Shiloh. His drum was smashed by a piece of shell, which occurrence won for him the appellation of "Johnnie Shiloh," as a title of distinction for his bravery. He was afterwards regularly mustered in and served also as a marker, and with his little musket so served on the battlefield of Chattanooga. At the close of that bloody day, the brigade in which he was was partly surrounded by rebels and was retreating, when he, being unable to fall back as fast as the rest of the line, was singled out by a rebel colonel, who rode up to him with the summons, scoundrel, "Halt! surrender, you —— little Yankee!" By way of order Johnnie halted, brought his piece to the position of charge bayonet, thus throwing the colonel off his guard. In another moment the piece was cocked, fired, and the colonel fell dead from his horse. Simultaneously with this the regiment was fired into, when Johnnie fell as though he had been shot, and laid there until darkness closed in, when he arose and made his way toward Chattanooga after the rest of the army. A few days later he was taken prisoner with others whilst detailed to bring up the supply trains from Bridgeport.

When he returned to service, General Thomas was in command of the army of the Cumberland. He received him with the warmest enthusiasm, made him an orderly sergeant, and attached him to his staff. At Chickamauga he was struck with a fragment of a shell in the hip, and at Atlanta, while he was in the act of delivering a despatch from General Thomas to General Logan, when a ball struck his pony obliquely near the top of his head, killing him and wounding his fearless little atom of a rider in the right ear.

For his heroic conduct he was made a sergeant by Rosecrans, who placed him upon the Roll of Honor, and attached him to the head-quarters of the army of the Cumberland, while a daughter of Chief-Justice Chase presented him with a silver medal inscribed, "Sergeant Johnnie Clem, Twenty-second Michigan Volunteer Infantry, from N. M. C.," which he worthily wears as a priceless badge of honor upon his left breast, in connection with his Grand Army medal.

Now (1890) Captain Clem is holding the important positions of Depot Quartermaster, Depot Commissary, ordnance office, Columbus, Ohio.

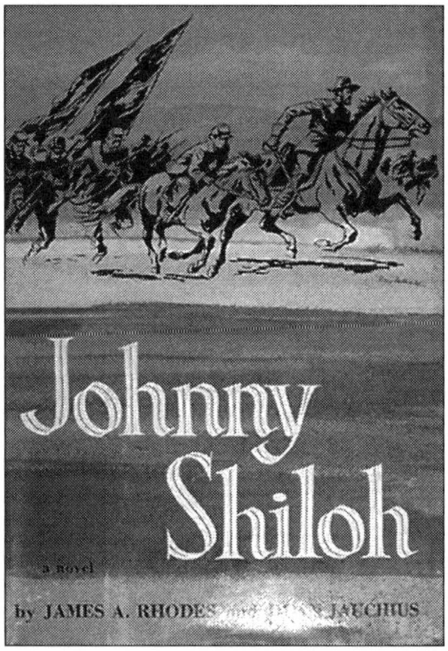

Above: Historical novel based on Clem's life. Co-authored in 1959 by James A. Rhodes, who would later serve four terms as Ohio's governor.

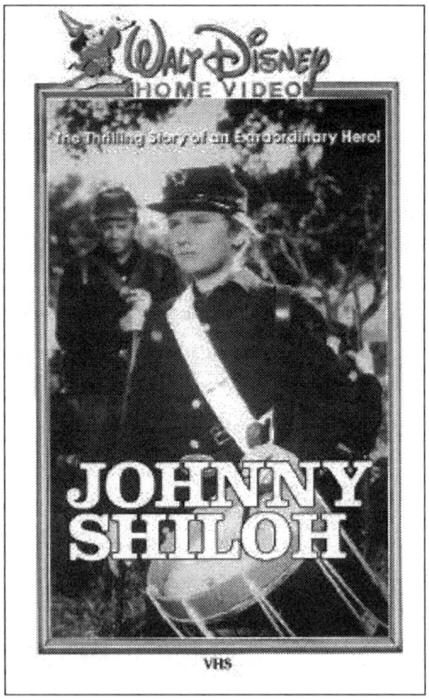

Above: The movie released in 1963 by Disney.

Right: Bronze statue of John Clem dedicated in 1999 at Sixth Street Park, Newark, Ohio.

Detail of plaque on base of statue.

Editor's Note:

The excerpts from Howe's history book on the previous pages quote John Clem's sister Lizzie as saying her brother Louis died with General Custer at the Battle of Little Big Horn on June 25, 1876. However, Louis' name does not seem to appear on the lists of soldiers or civilians killed in that battle. A transcript of a letter between John Clem and his parents dated May 23, 1876, indicates that Louis may have been killed at Custer City (Dakota Territory) on May 1, 1876 in some kind of conflict with Indians.*

*Catholic Record Society, Diocese of Columbus, Bulletin, Vol. 2, No. 7, July 1976.

Left: Monument dedicated to John Clem on the lawn of the Licking County Courthouse in Newark, Ohio.

Above: Movie based on Clem's life released in 2007 by Historical Productions, Inc., Cincinnati, Ohio. The film premiered in Newark.

Above: John Clem (left) and his brother, Louis, who was four years younger. John often wrote letters to Louis, offering brotherly advice, like in this excerpt from a letter written in December of 1872:

"Keep to your books. Read history at all spare times. It is both interesting and instructive. Be careful with your morals. Write home often and let them know how you are getting along."

Louis, who had long wanted to go out west, was killed in a fight with Indians in 1876 in Custer City, Dakota Territory. Louis is buried (or is at least listed on the Klem family headstone) in Mt. Calvary Cemetery on the south edge of Newark. *See photo opposite.*

Source: Catholic Record Society, Diocese of Columbus, Bulletin, Vol. 2, No. 7, July 1976.

Clem War Record on Trial

Historians disagree on some details of John Clem's military service. Two of the main points of contention are whether Clem actually participated in the Battle of Shiloh and whether he shot a Confederate colonel at the Battle of Chickamauga.

In the January 1989 issue of the magazine, *Civil War Times Illustrated*, Greg Pavelka, a park ranger in North Dakota, claimed that John Clem could not have been at the Battle of Shiloh. Some records indicate that Clem served with the 3rd Ohio, 22nd Michigan and 22nd Wisconsin. Apparently the 3rd Ohio did not participate in the Battle of Shiloh and the other two units were not organized until after the Battle of Shiloh.

It was also argued that records show no Confederate colonel being wounded at the Battle of Chickamauga.

In response to these claims a "trial" was held in Newark on October 14, 1989, to determine the validity of Clem's war record. Acting as council for the defense was Dr. R. Dean Jauchius, a professor at Franklin State University in Columbus and former Newark resident. Jauchius is co-author of the 1959 novel "Johnny Shiloh," which he wrote with James A. Rhodes, who would later become Ohio's governor. Jauchius is also a former Marine decorated for his action at Iwo Jima. The jury was made up of local politicians and civic leaders. Licking County Common Pleas Court Judge Neil M. Laughlin was foreman of the jury. Both sides were given 20 minutes to make their cases.

Jauchius contended that since Clem was only about nine years old when the Army took him in (below the legal age limit) his name would not have been listed on official regimental rosters in order to protect officers from being charged with allowing a fraudulent enlistment.

Jauchius also said that Clem had indeed shot a colonel at Chickamauga, and that the colonel, who later became an attorney in Texas, met and talked with Clem, exclaiming: "So you're the little (expletive) who shot me."

The jury voted on a verdict and cleared Clem of all charges that he may have padded his war record. When the verdict was announced, the courtroom crowd of about 350 people cheered.

Pavelka, discouraged at the verdict, lamented, "You can't fight with a legend."

General Dwight Beach, great-grandson of John Clem, was in attendance and agreed with the verdict. "That's the way the decision had to go. There are no living souls who know all the facts," he said.

Source: The Advocate, Newark Ohio, October 15, 1989.

Above: This boarded-up house at 26 W. Harrison Street in Newark, Ohio is thought to be John Clem's childhood home. If so, it is an important historical site, not only for Newark but for our national heritage.

Left: Klem headstone in Mt. Calvary Cemetery in Newark, is inscribed with the names of John Clem's brother Louis, his mother Magdalene, father, Roman and Roman's second wife, Elizabeth. John Clem is buried in Arlington National Cemetery in Washington D.C.

F
WAR

F
WAR
1812
C

MEXICAN
WAR
VETERAN

F
VETERAN
WAR
1861-1865
C

CUBA
SPANISH WAR VETERANS
ARMY
UNITED
NAVY
1898-1902
U.S.A.
PHILIPPINE ISLANDS
PORTO RICO

WORLD WAR
US

WORLD WAR TWO

KOREA

VIETNAM
U
S
1964-1975

PERSIAN GULF WAR
US
VETERAN

U.S.

WAR
US
VETERAN

AMERICA'S WARS

1775 to Present Day

The initial intent of this chapter was to provide comparative, county-level facts and figures for each war in which Licking County men and women were involved. We hoped to show how many Licking Countians served in each war, how many were killed, how many were wounded, etc. The idea was to provide a quick comparison of the relative impact each war had on the residents of our county.

However, facts and figures about wars are not as readily available as one might expect. Records and statistics from some time periods are difficult to locate, if they exist at all. There have also been fires and floods over the years in which even more recent records have been destroyed.

In addition, different kinds of information is scattered across a variety of federal, state and county offices. The facts and figures are often fragmentary and inconsistent.

There is also the issue of what constitutes a "Licking County Veteran." Do we count only men and women who were born in Licking County? Do we count those who officially entered the service in Licking County but actually lived elsewhere – or vice versa? Do we count those who may have only lived here briefly before, during or after their time in service?

A WORK IN PROGRESS

This book is by no means a complete or completely accurate record. This compilation of information is intended to serve the community and it will take the accumulated knowledge of the community to make it complete. If you have reliable information that may help correct errors on these pages, please contact the publisher at the address given in the front of the book.

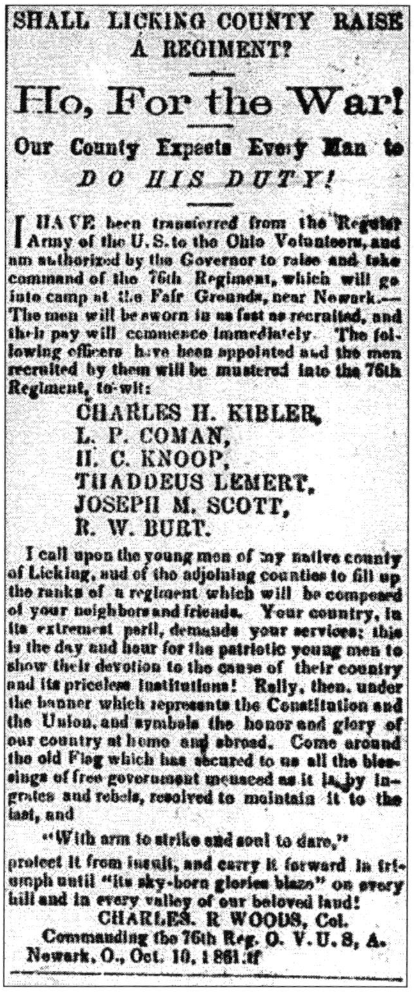

SHALL LICKING COUNTY RAISE A REGIMENT?

Ho, For the War!

Our County Expects Every Man to DO HIS DUTY!

I HAVE been transferred from the Regular Army of the U. S. to the Ohio Volunteers, and am authorized by the Governor to raise and take command of the 76th Regiment, which will go into camp at the Fair Grounds, near Newark.—The men will be sworn in as fast as recruited, and their pay will commence immediately. The following officers have been appointed and the men recruited by them will be mustered into the 76th Regiment, to-wit:

CHARLES H. KIBLER,
L. P. COMAN,
H. C. KNOOP,
THADDEUS LEMERT,
JOSEPH M. SCOTT,
R. W. BURT.

I call upon the young men of my native county of Licking, and of the adjoining counties to fill up the ranks of a regiment which will be composed of your neighbors and friends. Your country, in its extremest peril, demands your services; this is the day and hour for the patriotic young men to show their devotion to the cause of their country and its priceless institutions! Rally, then, under the banner which represents the Constitution and the Union, and symbols the honor and glory of our country at home and abroad. Come around the old Flag which has secured to us all the blessings of free government menaced as it is, by ingrates and rebels, resolved to maintain it to the last, and

"With arm to strike and soul to dare,"

protect it from insult, and carry it forward in triumph until "its sky-born glories blaze" on every hill and in every valley of our beloved land!

CHARLES R. WOODS, Col.
Commanding the 76th Reg. O. V. U. S. A.
Newark, O., Oct. 10, 1861. tf

Above: Advertisement in the Newark Advocate, Oct. 1861.

Opposite: Grave markers.

REVOLUTIONARY WAR
1775 – 1782

Synopsis of the War

The war for American independence between the Colonists and Great Britain. Also known as the American Revolution and the War of American Independence.

The Revolutionary War ended before Licking County officially existed. We do know that several Revolutionary War veterans lived in Licking County after the war and that several are buried here, but there is no way to list any accurate or meaningful statistics for how many "Licking County" men served in this war.

National Statistics

Source: Department of Veterans Affairs, Wash. DC

Total U.S. Service Members: 217,000

Battle Deaths: 4,435

Non-mortal Woundings: 6,188

Licking County Statistics

Sources: "Soldiers of the American Revolution Who Lived in the State of Ohio" and "The Official Roster: Soldiers of the American Revolution Buried in Ohio."

Lived in Licking County: 45

Buried in Licking County: 74

General Overview of the War

The path leading to the Revolutionary War was marked with mounting tensions between Colonial America and its British rulers. The tensions began ten years before the war with the implementation of the Stamp Tax in 1765. The Stamp Tax Act placed a tax on documents and newspapers, greatly frustrating the colonists. Colonists began to demonstrate against the tax burden imposed by Great Britain.

The most well known demonstration is the emblematic Boston Tea Party of December 16, 1763. After refusing to pay taxes to Great Britain for imported tea and after failed attempts to have the tea returned, more than 100 individuals threw approximately 45 tons of tea overboard from ships docked at Boston.

Tensions regarding taxation without representation continued to influence the relationship between the American Colonies and Great Britain. In 1775 at Lexington, Massachusetts, following Paul Revere's warning about the approaching British, the Revolutionary War began with a "shot heard 'round the world."

In 1776, the Declaration of Independence was penned and signed by several famous figures, including Benjamin Franklin, John Adams, John Hancock, Samuel Adams, Thomas Jefferson, and many others.

The war continued with George Washington serving as Commander in Chief of the Continental Army until the British forces surrendered on October 19, 1781, solidifying America's independence and ushering in a new era. The victory over Great Britain allowed colonists to establish their own government. Their efforts culminated with the adoption of the United States Constitution on September 17, 1787 in Philadelphia, Pennsylvania.

War of 1812
1812 – 1815

Synopsis of the War

A war between Great Britain and the United States that has also been referred to as the second American war for independence.

"Licking county was rapidly filling up with settlers from the eastern States when war was declared against Great Britain in 1812, and the young men were called upon to drop their axes and go forth to protect the frontiers against the wily savage and his British master."

From: *The History of Licking County O.* by N. N. Hill, Jr., 1881.

According to N. N. Hill Jr., Licking County contributed at least four companies and a company of cavalry to the cause. In response to a request for three volunteer regiments by Ohio's Governor Meigs, Lewis Cass was chosen colonel of the third regiment. He was joined by the Licking County volunteers at Findlay. On July 11th, they crossed into Canada, Colonel Cass claiming the honor of being the first man in the war of 1812 to have stepped in arms upon British soil.

National Statistics

Source: Department of Veterans Affairs, Wash. DC

Total U.S. Service Members: 286,730

Battle Deaths: 2,260

Non-mortal Woundings: 4,505

Ohio Statistics

Source: Ohio Historical Society

Ohio provided 1,759 officers and 24,521 enlisted men

Licking County Statistics

No reliable statistics located.

General Overview of the War

An often forgotten war, The War of 1812 once again poised the young nation of the United States in opposition to Great Britain. Under the command of President Thomas Jefferson, the U.S. participated in international trade but attempted to remain neutral in foreign affairs and conflicts.

However, in the early 19th Century, Napoleonic France and Great Britain were entrenched in a conflict. France and Great Britain eventually enacted legislation to prohibit U.S. vessels from trading with European countries. The prohibition on U.S. trade was largely unsuccessful but Great Britain continued to search U.S. ships when in European waters, often to force and coerce U.S. seamen to serve in the Navy of Great Britain. United States officials sought a declaration of war against Great Britain in 1812.

In 1813 the war intensified along Ohio's North Coast, off the shores of Lake Erie. In the early morning hours of September 10, 1813, Oliver Hazard Perry, Commodore of the U.S. fleet on Lake Erie, and his men encountered six vessels belonging to Great Britain's Royal Navy, resulting in the Battle of Lake Erie.

The Battle of Lake Erie consisted of six British vessels commanding 63 cannons with only 54 cannons throughout the American nine-vessel fleet. It is considered one of the most influential battles of the war, securing Lake Erie and forcing a British retreat. Perry's victory was later considered to be a contributing factor in ensuring the security of the Ohio and Michigan territories. Following the intense battle, Perry penned a letter to General Harrison that contains the now famous quote, "We have met the enemy and they are ours" in reference to his success in the battle.

In 1814, Great Britain and the U.S. signed the Treaty of Ghent, ending the war.

Indian Wars
1817 – 1898 (approx.)

Synopsis of the War

A series of conflicts between the colonial or federal government and the Indians (indigenous peoples) of North America.

"Of the Indian War but little can be said, as the first settlement in Licking County was made after the Greenville treaty, and our pioneers were not molested by the savages of the forest within the limits of the county."

There were isolated incidents with Indians in Licking County. One account tells of a night in April of 1800 when two Indians stole the horses of settlers Hughes and Ratliff from a little enclosure near their cabins. The men pursued the Indians, eventually surprising them at their camp the next morning where they shot them dead. "They returned to their cabins with the horses and 'plunder' taken from the Indians, and swore mutual secrecy for this violation of law."

From: *The History of Licking County O.* by N. N. Hill, Jr., 1881.

National Statistics

Source: Department of Veterans Affairs, Wash. DC

Total U.S. Service Members: 106,000 (estimate)

Battle Deaths: 1,000 (estimate)

General Overview of the Wars

During the 19th century, the United States' interest in expanding its territory met opposition from Native Americans. The series of conflicts are now referred to collectively as the Indian Wars.

The battles that raged on, in chronological order, included the Battle of Tippecanoe in Indiana; the Creek War of Georgia and Alabama; three separate Seminole Wars in Florida; the Black Hawk War in Illinois and Wisconsin; the Sioux Wars of Wyoming, Minnesota, and South Dakota; Apache Attacks in New Mexico, Arizona, Texas, and Mexico; the Modoc War of California and Oregon; the Red River War in Texas; the Battle of the Rosebud at Rosebud Creek in Montana; the Battle of Little Big Horn in Montana; the Nez Perce War of Oregon, Idaho, and Montana; and the Wounded Knee Massacre in South Dakota.

Tecumseh, a well known Shawnee warrior from Ohio, collaborated with tribes across the country to fight for the rights of the tribes. Tecumseh and his followers joined forces with the British during the War of 1812, hoping the British would help the tribes regain their homeland.

In 1876, Ohioan, General George A. Custer was killed when he and 250 soldiers clashed with Sioux warriors at the Battle of Little Big Horn.

Although the Indians were victorious over Custer, the tribes suffered greatly during the entire 19th Century and previous eras as a result of the policies and actions of settlers from European countries. The result of battles, constant struggle for land, overt maltreatment by colonists, exposure to novel diseases, limited access to resources, and outright acts of genocide led to the decimation of numerous Indian tribes and respective cultures – a cost of war that cannot be measured.

MEXICAN WAR
1846 – 1848

Synopsis of the War

An armed conflict between the United States and Mexico over the ownership of the territory encompassing New Mexico and California.

"Two infantry companies and one of cavalry were almost wholly recruited in this county..."
From: *The History of Licking County O.* by N. N. Hill, Jr., 1881.

According to N. N. Hill, Jr., The cavalry, known as "Duncan's Mounted Rangers," numbered one hundred men besides the officers, and nearly every member furnished his own horse. The company left Newark with "considerable enthusiasm," in May of 1847, riding all the way to Cincinnati. Here they went aboard the steamer "Star Spangled Banner," and in the latter part of June arrived in New Orleans and then on to Seralvo on the Rio Grande.

National Statistics
Source: Department of Veterans Affairs, Wash. DC

Total U.S. Service Members: 78,718

Battle Deaths: 1,733

Other Deaths (in theater): 11,550

Non-mortal Woundings: 4,152

Ohio Statistics
Source: Ohio Historical Society

Enlisted: 7,000 (estimate)

Licking County Statistics
Source: Various

Battle Deaths: 3

Other Deaths: 5

Non-mortal Woundings: 1

General Overview of the War

By the spring of 1846, the United States was once again a nation at war. The 11th president of the U.S., James Polk was interested in expanding the country's territory.

Polk became increasingly interested in the territory encompassing New Mexico and California. After a failed attempt to purchase the territory from Mexico, Polk ordered General Zachary Taylor and his troops to station themselves on the banks of the Rio Grande. Taylor's maneuver was perceived by Mexican troops as an act of aggression, provoking them to attack.

President Polk told congress that "American blood had been shed on American soil," urging them to declare war. In May of 1846, Polk's persuasion paid off and congress declared war against Mexico.

United States' forces proved victorious in battle after battle, eventually leading them to the occupation of Mexico City. In 1848, Mexico relented and provided the U.S. with New Mexico and California territory in exchange for money.

The acquisition of this territory greatly increased the geographic size of the United States.

Civil War
1861 – 1865

Synopsis of the War

The United States' war between the northern and southern states over states' rights and slavery issues. Also known as the War Between the States, the War of the Rebellion, War of Succession and War of Southern Independence.

The federal enrollment of men in Licking County, subject to military duty for the years 1863-64, was 4,880. The county furnished 3,160 volunteers to serve out of state, 368 men who were drafted and 404 who served in the state in defense of Cincinnati.

From: *The History of Licking County O.* by N. N. Hill, Jr., 1881.

National Statistics

Source: Department of Veterans Affairs, Wash. DC

Union:

Total U.S. Service Members: 2,213,363
Battle Deaths: 140,414
Other Deaths (in theater): 224,097
Non-mortal Woundings: 281,881

Confederate:

Total Service Members: 1,050,000
Battle Deaths: 74,524
Other Deaths (in theater): 59,297

Ohio Statistics

Source: History of Ohio by J.P. Lawyer, Jr., 1904

Total Service Members: 340,000
Battle Deaths: 6,536
Mortal Woundings: 4,674
Disease & Other Deaths: 13,354

Licking County Statistics

Source: Various

Total Service Members: 3,932
Deaths: 517

General Overview of the War

In 1861 Abraham Lincoln, known as an opponent of slavery, became the 16th president of the United States. Lincoln's election was perceived as a threat to southern states and several of those states began seceding from the union. The secession started with South Carolina and eventually Mississippi, Florida, Alabama, Georgia, Louisiana, Texas, Virginia, Arkansas, Tennessee, and North Carolina followed.

Shortly after secessions began, representatives from the several of the seceding states met in Montgomery, Alabama to create the Confederate Constitution. Jefferson Davis was appointed president of the Confederacy *pro tempore*. The Confederacy and its troops moved quickly to secure federal forts.

The Civil War began on April 12, 1861 at Fort Sumter when Confederate troops fired on the fort held by Union soldiers. Fort Sumter, with limited supplies, fell to the Confederacy. With a decisive Confederate victory, additional states joined the Confederacy.

Battles of the Civil War raged on for months that turned into years, with occasional victories for both sides. After numerous battles, each side was beginning to show signs of weakness. In early spring of 1865, Ohio-born General William Tecumseh Sherman and his troops marched through the south leaving a fiery trail of destruction in their wake. During the march, Sherman took control of Atlanta and continued on through Georgia and the Carolinas, severely undermining the Confederacy.

In the spring of 1865, the Confederacy fell and surrendered to Union forces.

Spanish-American War
1898 – 1902

Synopsis of the War

The war between The United States and Spain, arising out of Spanish policies in Cuba.

According to the "History of Licking County," by E.M.P. Brister, there were three organized military companies in Licking County when the war with Spain was declared. Two were based in Newark, one in Hebron.

"Of both infantry and artillery furnished by Licking County, it may be said without boasting that they were excelled by none with whom they were compared in the field, whether from Ohio or from our sister states."

"In addition to the organized companies, a number of Licking County boys entered the regular army, and also joined other state organizations, outside the county, wherever they could find places." *(Quotes from Brister)*

National Statistics
Source: Department of Veterans Affairs, Wash. DC

Total U.S. Service Members: 306,760

Battle Deaths: 385

Other Deaths in Service (non-theater): 2,061

Non-mortal Woundings: 1,662

Ohio Statistics
Source: Ohio Historical Society

Soldiers from Ohio: 15,300

Deaths: 230

Licking County Statistics
Source: Various

Total Service Members: 200 (estimate)

General Overview of the War

The United States issued a formal declaration of war on Spain on April 25, 1898. The Spanish-American War was precipitated by increasing tensions regarding Cuba's liberation from Spanish control. The tensions were exacerbated by an explosion that sank the USS Maine, an American warship, off the coast of Cuba. The explosion was attributed to an attack by Spain. However, the exact cause of the explosion has been the subject of much speculation over the years.

At the onset of the war, Cuba was fighting for independence from Spain. U.S. forces under the command of President William McKinley were effective in defeating Spanish forces. Future president Theodore Roosevelt led the Rough Riders in battle, securing San Juan Hill.

Founder of the American Red Cross, abolitionist, humanitarian, and a nurse on the battlefields of the Civil War, Clara Barton directed the Red Cross aid to American soldiers, prisoners of war, and refugees. The Spanish-American War marked the first time the Red Cross served the armed forces in this manner, starting a relationship that continues today.

The Spanish-American War served as an entrée into the role of world power for the United States. After a successful campaign by U.S. troops, the Treaty of Paris was signed, acknowledging the U.S. victory and removing Spain from some of its territories, including Cuba, Guam, Puerto Rico, and the Philippines.

The Mexican Campaign

Between the Spanish-American War and WWI, there were two occasions (in 1914 and again in 1916) that required sending American forces to Mexico for a period of months. It was during the 1914 engagement that Newark's Harry C. Beasley served on the USS Florida, earning a Medal of Honor.

World War I
1914 – 1918

Synopsis of the War

Great Britain, France, Russia, Belgium, Italy, Japan, the United States, and other allies defeated Germany, Austria-Hungary, Turkey, and Bulgaria. Also known as the Great War.

National Statistics
Source: Department of Veterans Affairs, Wash. DC

Total U.S. Service Members: 4,734,991

Battle Deaths: 53,402

Other Deaths in Service (non-theater): 63,114

Non-mortal Woundings: 204,002

Ohio Statistics
Source: Ohio Historical Society

Soldiers from Ohio: 200,000

Deaths: 6,500

Licking County Statistics
Source: Various

Total Service Members: 2,596

Deaths: 82

General Overview of the War

The war we now call World War I started in July of 1914 when Austria-Hungary declared war on Serbia. Eventually the conflict evolved and spread to involve Germany, Russia, Great Britain, France, and other nations.

The United States entered the war in 1917 when it declared war on Germany following attacks on United States' warships in the Atlantic. The perils of war accompanied by a deadly influenza outbreak wore on troops from both sides and by late 1918 World War I came to an end. The last armistice was signed on November 11, 1918. Austria-Hungary was broken into smaller countries. The Treaty of Versailles punished Germany severely through economic sanctions and loss of territory.

World War I has been credited with setting the tone for many events of the 20th Century, including World War II.

World War II
1941 – 1945

Synopsis of the War

Great Britain, France, the Soviet Union, the United States, China, and other allies defeated Germany, Italy, and Japan.

National Statistics

Source: Department of Veterans Affairs, Wash. DC

Total U.S. Service Members: 16,112,566

Battle Deaths: 291,557

Other Deaths in Service (non-theater): 113,842

Non-mortal Woundings: 671,846

Ohio Statistics

Source: Ohio Historical Society

Soldiers from Ohio: 839,000

Deaths or missing: 23,000

Licking County Statistics

Source: Various

Deaths: 194

General Overview of the War

Adolf Hitler's Germany ushered in the start of World War II in 1939 when German troops invaded Poland.

Britain and France were quick to declare war on Germany, while the United States remained neutral. Germany invaded and took several additional countries, including Denmark, Norway, the Netherlands, and France.

In 1940, Germany attempted to take Britain through rapid air strikes. Britain's Royal Air Force was able to defend the country and repel Germany. Italy, as an ally with Germany, expanded the war front by invading Greece and countries in northern Africa.

In 1941, Japan, at war with China for several years, launched an unexpected attacked on Pearl Harbor, a U.S. Naval Base in Hawaii. The attack rendered the U.S. significantly weaker militarily and led to the United States' entry into World War II.

The hostility resulting from Japan's surprise attack carried U.S. forces to victory at the Battle of Midway in June of 1942, where Japan suffered significant losses.

In August of 1945, for the first and only time, two atomic bombs were used in battle. The United States dropped an atomic bomb on the Japanese cities of Hiroshima and Nagasaki. The Hiroshima bomb was dropped from a plane piloted by Paul Tibbet of Ohio. It is estimated that 140,000 people died as a result of these bombs. An estimated 20,000 were soldiers.

Hitler committed suicide in 1945, following the death of Mussolini and the downturn of his empire. The events of 1945 set the stage for the end of World War II.

By December 31, 1946 the Axis powers (Germany, Japan, and Italy) had surrendered and President Truman officially proclaimed the cessation of hostility in World War II.

Korean War
1950 – 1953

Synopsis of the War

A conflict between North Korea, aided by China, and South Korea, aided by United Nations forces consisting primarily of U.S. troops.

National Statistics
Source: Department of Veterans Affairs, Wash. DC

Total U.S. Service Members: 5,720,000

Battle Deaths: 33,739

Non-mortal Woundings: 103,284

Ohio Statistics
Source: Ohio Historical Society

Deaths: 1,777

Wounded: 4,837

Licking County Statistics
Source: Various

Deaths: 46

General Overview of the War

Resulting directly from the Allied victory in World War II, the Japanese occupation of the Korean peninsula ended in 1945. During the tumultuous transition, Korea was divided into two sections. With a border established at the 38th parallel, North Korea embraced communism while South Korea did not. The division often separated families once unified in a single country. The division also set the stage for what became the Korean War.

In 1950, North Korean troops crossed the 38th parallel and aggressively attacked South Korea. North Korea was heavily armed with tanks and other armaments from the Soviet Union, allowing them to quickly overrun South Korea's defenses.

The United States strongly opposed communist North Korea's attack on South Korea. In June of 1950, the U.S. deployed troops to the Korean peninsula to aid South Korea. As a world power, and at a time the U.S. was interested in the containment of communism, the U.S. with its military superiority expected to quickly win the conflict in Korea. However the battles raged on and the Korean War lasted more than three years. United States' forces suffered tens of thousands of deaths.

The Korean War resulted in an armistice in 1953 after years of struggle over rights to territory on the Korean peninsula. Today the peninsula remains divided into two regions, still often in opposition with each other.

VIETNAM WAR
1965 – 1975

Synopsis of the War

A protracted military conflict between the Communist forces of North Vietnam supported by China and the Soviet Union and the non-Communist forces of South Vietnam supported by the United States.

National Statistics
Source: Department of Veterans Affairs, Wash. DC

Total U.S. Service Members: 8,744,000

Deployed to Southeast Asia: 3,403,000

Battle Deaths: 47,434

Other Deaths (in theater): 10,786

Other Deaths in Service (non-theater): 32,000

Non-mortal Woundings: 153,303

Ohio Statistics
Source: Ohio Historical Society

Deaths: 2,997

Wounded: 20,000

Licking County Statistics
Source: Various

Deaths: 38

General Overview of the War

In 1965, combat units were deployed to Vietnam beginning the United States' involvement in the Vietnam War. The United States again found itself defending the south, non-Communist portion of a divided county against its northern, communist counterpart.

The 1954 Geneva Accords divided the country into two regions at the 17th parallel, a decision that was intended to last only two years. The involvement of U.S. troops escalated in 1968 following the operation of North Vietnam's Tet Offensive. North Vietnam launched the Tet Offensive, an attack on most major South Vietnam cities, in an attempt to win the war in one step.

However, South Vietnam's defenses in collaboration with American troops were able to withstand the Tet Offensive. The Tet Offensive showed that the North Vietnam Communist forces were not as weak as once thought. It also provided a glimpse for American public into the horrors of war with extensive media coverage. Although an American victory, the Tet Offensive was costly in terms of American lives.

The antiwar movement began to grow and gain momentum among the American public. Protests about the government's use of the draft and the U.S. involvement in the Vietnam conflict were commonplace. Ohio National Guard troops opened fire on a crowd of protestors on the campus of Kent State University on May 4, 1970, killing four students.

In January of 1973, The United States and North Vietnam entered into a cease-fire. All U.S. military left Vietnam by that March. The conflict between North and South Vietnam continued until 1975 when Saigon, the capital of South Vietnam, fell to North Vietnam. North Vietnam reunified the country as the Socialist Republic of Vietnam.

Persian Gulf War
1990 – 1991

Synopsis of the War

Armed conflicts between Iraq and a coalition of nations including the United States and Britain in response to Iraq's invasion of Kuwait.

National Statistics
Source: Department of Veterans Affairs, Wash. DC

Total U.S. Service Members: 2,322,000

Deployed to Gulf: 694,550

Battle Deaths: 147

Other Deaths (in theater): 235

Other Deaths in Service (non-theater): 1,590

Non-mortal Woundings: 467

Ohio Statistics
Source: Ohio Historical Society

Deaths: 16

Licking County Statistics
Source: Various

Deaths: 1

General Overview of the War

Prompted by an August 2, 1990 invasion of Kuwait by the Iraqi Republican Guard, the United States launched operation Desert Shield on August 7, 1990 to prevent Iraqi forces from expanding their offensive into Saudi Arabia.

The United Nations joined the United States by condemning Iraq's occupation of Kuwait. The UN issued an ultimatum instructing Iraq's dictator Saddam Hussein to vacate Kuwait or face repercussions in the form of U.S.-led military action from a coalition of forces.

On January 17, 1991, two days after the deadline set by the UN's ultimatum, U.S.-led coalition forces began operation Desert Storm, conducting air strikes on Iraqi positions. Ground forces began their attacks on Iraqi interests on February 24, 1991.

The U.S. declared Kuwait City liberated on February 27, 1991. Coalition forces had secured positions deep inside Iraq, prompting President George H.W. Bush to suspend military action in the region. The following day a cease-fire agreement was enacted.

The conflict is also referred to as the First Gulf War and the First Iraq War. It is also often referred to Desert Shield-Desert Storm in recognition of the two major operations of the war.

THE WAR IN IRAQ & AFGHANISTAN
OCT. 2001 – PRESENT

Synopsis of the War

The United States involvement in armed conflicts in Afghanistan and Iraq in an attempt to disrupt the activities of international terrorist cells.

National Statistics
Source: Department of Veterans Affairs, Wash. DC

Figures as of Aug. 2009

Hostile Deaths: 4,008

Non-hostile Deaths: 1,109

Wounded in Action: 35,205

Ohio Statistics
Source: The Washington Post

Figures as of Oct. 6, 2009

Afghanistan Deaths: 23

Iraq Deaths: 184

Licking County Statistics
Source: Various

Figures as of Oct. 6, 2009

Iraq Deaths: 2

General Overview of the War

The conflict sometimes referred to as the "Global War on Terror," was prompted by the September 11, 2001 terrorist attacks on the United States. The conflict is also referred to as the Overseas Contingency Operation.

The objectives of the war for the U.S. are to protect the citizens of the United States and its allies, to protect the business interests of the U.S. and allies at home and abroad, break up terrorist cells in the U.S., and disrupt the activities of the international network of terrorist organizations made up of a number of groups under the umbrella of al-Qaeda.

This conflict is ongoing.

Afghanistan

American and British forces launched Operation Enduring Freedom against Taliban and al-Qaeda forces in Afghanistan in October 2001. Following military intelligence, the United States' mission is to find Bin Laden, destroy the Taliban, and promote stability in the region. Although the U.S. was quickly successful in removing the Taliban from control, as of this writing, the war in Afghanistan continues and recent events have included Taliban-led insurgency. Osama Bin Laden remains at large and continues to be hunted by U.S. forces.

Iraq

In March of 2003, American and British- led coalition forces entered Iraq, launching Operation Iraqi Freedom. The conflict began after Iraq was suspected of amassing weapons of mass destruction and funneling money to support terrorist organizations. Iraq's dictator, Saddam Hussein was removed from power and executed for crimes against the nation's citizens. American troops continue to work with the new Iraqi government to enhance stability in the country.

THE FALLEN

Licking County Men Who Died in Service

The list on the following pages, courtesy of The Licking County Veterans' Commission, begins with the most recent casualties and continues back through the Mexican War. The list is thought to be accurate from the Vietnam War forward. For wars prior to Vietnam, it is likely that there are missing names or discrepancies. In the Civil War, for example, many men lived through the war but died sometime later from the wounds or disease they acquired in the war. Depending on the length of time between the end of the war and his death, a veteran may or may not show up on this list as being a war casualty.

Compiling a complete and accurate list of every person from Licking County who ever died while in the military or even a list of all those who served, would be an impossible task. Too much time has passed, records have been lost, discrepancies exist, and record-keeping has not always been a priority.

Although today's computers give us the advantage of searching certain records faster than earlier historians could ever have imagined, the records themselves are not always complete or accurate.

The following excerpt from N. N. Hill, Jr.'s "History of Licking County Ohio," published in 1881, echoes this age-old problem.

"Would that these pages could present in letters of gold the names of Revolutionary patriots who contributed so much to make Licking County what it is, but it has been impossible to procure a full list!"

Above: The headstone of a Civil War veteran, in the Calvanistic Welsh Church Cemetery on Sharon Valley Road.

Opposite: Soldier's grave in the old veterans cemetery in Johnstown, Ohio.

Editor's note: The photos of veterans' grave markers on the following pages are random images from around the county. Not all of these veterans died in service.

Above: Lance Corporal Devon Seymour of Saint Loiuisville, killed in action in Iraq in 2005.

Devon was a very patriotic young man. He tried to enlist in the Marine Corps after the terrorist attacks of 9/11 but was turned down because he was only 17. After he turned 18, he was able to enlist and was eventually deployed to Iraq as a gunner on a Humvee (jeep).

On June 9, 2005, he was killed along with four other soldiers when an explosion destroyed their Humvee.

Below: Memorial to Devon Seymour in the Saint Louisville Village Park.

Last Name .... First Name .... City .... Branch .... War

Iraq War

Last Name	First Name	City	Branch	War
Seymour	Devon	St. Louisville	USMC	Iraq
Nice	Joseph	Newark	USMC	Iraq

Gulf War

Last Name	First Name	City	Branch	War
Danielson	Donald	Newark	Army	Desert Storm

Vietnam

Last Name	First Name	City	Branch	War
Bond	George A.	Newark	USMC	Vietnam
Bragg	Jr. Fred G.	Etna	Army	Vietnam
Brown	Jackie R.	Newark	USMC	Vietnam
Bussey	Marvin W.	Jacksontown	Army	Vietnam
Crothers	Danny K.	Newark	USMC	Vietnam
Decker	Jr. William T.	Newark	Army	Vietnam
Edris	Richard J	Newark	Navy	Vietnam
Elswick	Robert W.	Newark	Army	Vietnam
Eppley	Gerald V.	Newark	USMC	Vietnam
Ford	Jr. Kenneth A.	Newark	USMC	Vietnam
Garven	Charles D.	Newark	Army	Vietnam
Holler	Carl W.	Utica	Army	Vietnam
Hyatt	Jerald M.	Newark	Army	Vietnam
Irving	Lee	Heath	USMC	Vietnam
Johnson	Ray E.	Alexandria	Army	Vietnam
Kean	Billie O.	Alexandria	USMC	Vietnam
Keeran	Willard D.	Newark	Army	Vietnam
Kinnard	Daniel L.	Newark	Army	Vietnam
Lamp	Jr. Arnold W.	Hebron	A.F.	Vietnam
Maze	David L.	Pataskala	Army	Vietnam
McIntyre	David A.	Newark	Army	Vietnam
Miller	James G.	Newark	Army	Vietnam
Myers	Paul R.	Newark	Army	Vietnam
Oliver	Walter B	Newark	Army	Vietnam
Rhodes	Grant A.	Newark	Army	Vietnam
Ritchey	Gary W.	Newark	Army	Vietnam
Ryan	Samuel F.	Utica	Army	Vietnam
Schmidt	Joseph	Newark	Army	Vietnam

Sens	Philip M.	Newark	USMC	Vietnam
Simpson	Douglas E.	Heath	Army	Vietnam
Smoke	Bruce A.	Newark	Army	Vietnam
Speaks	Paul E.	Newark	USMC	Vietnam
Stapleton	Clifford	Newark	Army	Vietnam
Weidner	Richard D.			Vietnam
Wick	Richard G.	Heath		Vietnam
Wood	William L.	Newark	Army	Vietnam
Yost	Jr, Howard E.	Newark	Army	Vietnam
Varner	Douglas	Newark		Vietnam

Korea

Beever	Franklin	Utica		Korea
Blade	Russell E.	Hebron	Army	Korea
Camp	Harold L.	Granville		Korea
Dobson	Carl L.	Newark		Korea
Elliott	Fred L.	Newark		Korea
Evans	Jr., Bryant	Newark	Army	Korea
Evans	Halbert Knapp	Granville	USN	Korea
Henderson	Roy A.	Newark	Army	Korea
Hickman	William G.	Newark	Army	Korea
Holler	Ivan Lowell	Utica	USMC	Korea
Holler	Walter J.	Newark		Korea
Jones	John E.	Newark	Army	Korea
King	Kenneth J.	Newark		Korea
Lantz	Dale E.	Granville		Korea
Lewis	Vaughn D.	Utica		Korea
Lyons	James R.	Hebron	Army	Korea
Mariani	James E.	Newark		Korea
Martin	James F.	Newark		Korea
McKee	Paul L.	Frazeysburg		Korea
Miller	Alfred D.	Newark		Korea
Neff	Robert M.	Pataskala		Korea
Osbourne	Warren L.	Newark		Korea
Peters	Ralph E.	Newark	Army	Korea
Phelan	William F.	Newark		Korea
Proctor	Charles E.	Utica		Korea
Ralston	Robert E.	Pataskala		Korea
Rardon	Milon E.		Army	Korea
Richards	Leonard Lee	Toboso	Army	Korea
Richardson	Joe E.	Newark		Korea
Riffle	John K.	Homer	Army	Korea
Rittenhouse	Henry J.	Newark		Korea

Note to Readers

This list is by no means complete or completely accurate. Readers who find errors or missing names are encouraged to contact the Licking County Veterans Commission office and the publisher of this book so that the list can be updated. A mailing address for the publisher can be found on page **9**.

Above: This headstone in the Fredonia Cemetery marks the grave of William G. Hickman, killed in the Korean War.

Below: Helen S. Rollins served as a nurse during WWII, Korea and Vietnam. She lies in South Lawn Cemetery in Utica.

Above: The military marker of Franklin D. Howard, veteran of the Korean War, in the Jacksontown Cemetery on Rt. 13 north of Jacksontown.

Below: Revolutionary War veteran, George Avery, Jr. is buried in the Old Colony Burying Ground in Granville.

Shaner	Edwin L.	Newark		Korea
Shaw	Raymond G.	Newark		Korea
Siegmund	Earl V.	Newark	Army	Korea
Small	Roscoe D.	Newark		Korea
Smith	James E.	Pataskala	Army	Korea
Snedden	Lawrence Austin	Newark	Army	Korea
Squiggins	George F.	Newark	Army	Korea
Swint	Carl E.	Buckeye-Lake		Korea
Tripp	Richard E.	Newark		Korea
Tucker	Stanley R.	Newark	Army	Korea
Varner	Alan L.	Newark		Korea
White	Carl W.	Newark		Korea
Wilfong	Roy L.	St. Louisville		Korea
Woods	Earl E.	Johnstown	Army	Korea
Wright	Roy N.	Newark	Army	Korea

World War II

Adkins	Richard Carl	
Baughman	Delmer D.	World War II
Becker	Jr., Peter	World War II
Bending	Robert L.	World War II
Benner	Benjamin C.	World War II
Beever	Russel	World War II
Beever	Spencer	World War II
Bickel	William F.	World War II
Blade	John C.	World War II
Booher	James R.	World War II
Bonnett	Chester H.	World War II
Boring	Joseph Wilbur	World War II
Bostwick	Herbert C.	World War II
Brannon	Robert G.	World War II
Britton	Russel P.	World War II
Brown	William R.	World War II
Brune	Henry W.	World War II
Burden	Robert F.	World War II
Burger	Leroy	World War II
Chopson	George	World War II
Clark	Amos B.	World War II
Chopson	George E.	World War II
Chrisman	James R.	World War II
Clippinger	Donald K.	World War II
Cochran	Donald Ivan	World War II
Conard	Ray Hamlin	World War II

Copeland	Robert	World War II
Crawmer	Curtis	World War II
Dankmer	Joseph L.	World War II
Davis	Clyde V.	World War II
Davis	Kenneth C.	World War II
Davis	Murle	World War II
Davis	Walter E.	World War II
DeCrow	Joseph C.	World War II
Denton	John J.	World War II
Dietzel	Charles T.	World War II
Dixon	John A.	World War II
Doneff	John L.	World War II
Duncan	Edward L.	World War II
Duncan	Olin M.	World War II
Durkin	Shirley L.	World War II
Emmerson	William L.	World War II
Eppley	George E.	World War II
Faller	Murray C.	World War II
Farquhar	Floyd F.	World War II
Flynn	Edmund A.	World War II
Fox	Richard C.	World War II
Fox	Robert L.	World War II
Fulton	John W.	World War II
Geiger	Byron W.	World War II
Geiger	Marion S.	World War II
Ghiloni	John P.	World War II
Gorey	William M.	World War II
Gough	Austin C.	World War II
Grady	John Patrick	World War II
Green	Charles W.	World War II
Greenlee	George	World War II
Greenwald	Robert L.	World War II
Griffith	Harry	World War II
Griffith	Robert E.	World War II
Gumm	James J.	World War II
Hannahs	Robert E.	World War II
Harper	Raymond T.	World War II
Harrington	Ralph C.	World War II
Haugh	Rex D.	World War II
Heinlein	Calvin S.	World War II
Hinger	Willard Joseph	World War II
Hitchcock	Raymond F.	World War II
Hoover	John W.	World War II
Hoagwood	George H.	World War II
Horn	Melvin F.	World War II

Above: Wilbur L. Bullock and Carl W. Bullock lie side-by-side in Evans Cemetery near St. Louisville. Both men served in the Army during WWII.

Below: Conrad S. Roberts served in the Navy during WWII and Korea. He lies in the Fredonia Cemetery.

Above: Veteran, Paul Swartz served as a Private in Co. D, 106 Infantry during WWII. He lies in Jacksontown Cemetery.

Below: Vernon C. Weaver served as a Pvt. in the 112 Med. Regt., 37 Div., in WWII. He lies in Kirkersville Cemetery.

Huesmann	Joseph L.	World War II
Hupp	Carroll L.	World War II
Inlow	Burdett S	World War II
Keaser	Paul E.	World War II
Kinmey	Gene S.	World War II
Krumm	Nelson F.	World War II
Layman, Jr.	Mark D.	World War II
LeBlanc	Dale R.	World War II
Leckly	Richard L.	World War II
Lee	Harold E.	World War II
Link	Louis W.	World War II
Litten	Warren N.	World War II
Loar	Paul H.	World War II
Luther	Emmett Paul	World War II
Lyle	Lested E.	World War II
Maniace	William	World War II
Martin	Charles W.	World War II
Martin	Richard L.	World War II
Mashcoff	Ralph E.	World War II
Mead	Carl J.	World War II
Mentzer	Ralph A	World War II
Meyer	Harold W.	World War II
Miller	Donald L	World War II
Miller	Ralph P.	World War II
Mitchell	Robert H.	World War II
Moore	Dorsey E.	World War II
Morganti	Cammilo J.	World War II
Murphy	Charles R.	World War II
Murphy	James E.	World War II
Murphy	Robert Dean	World War II
Murray	William E.	World War II
Myers	Leroy J.	World War II
McFarland	Andrew	World War II
Needles	Frederick L.	World War II
Negele	Josepy H.	World War II
Nelson	George Jack	World War II
Nolan	Earl Thomas	World War II
Nornan	Amos A.	World War II
Odell	Frederick J.	World War II
Oliver	William F.	World War II
Obannon	Thomas B.	World War II
O'Dell	Fred	World War II
Painter	Joseph R.	World War II
Parkinson	Raymond	World War II
Payne	Bradly L.	World War II

Penick	Ralph	World War II
Perfect	Truman A	World War II
Perfect	Jr., John D.	World War II
Perkins	John D.	World War II
Perkkins	William L.	World War II
Pettit	Kenneth L.	World War II
Piccirilli	Angelo C.	World War II
Pierce	Richard F.	World War II
Piper	Clell S.	World War II
Porterfield	John A	World War II
Priest	Walter D.	World War II
Pyle	Paul L.	World War II
Quick	Robert P.	World War II
Radcliff	Merton G.	World War II
Ramey	Austin M.	World War II
Ratliff	Wilfred A.	World War II
Reid	Edwin R.	World War II
Reid	Thomas H.	World War II
Rey	Leo P.	World War II
Reynolds	Harold Leroy	World War II
Ricciardo	Joseph A.	World War II
Richards	William M.	World War II
Richardson	John D.	World War II
Ricketts	John W.	World War II
Riley	Charles A.	World War II
Rine	Archie Ernest	World War II
Rittenhousse	Stanley	World War II
Rogers	Robert	World War II
Rusk	Floyd	World War II
Roike	Thomas	World War II
Rusk	Floyd D	World War II
Sager	Paul D.	World War II
Scalone	Joseph M.	World War II
Schmitt	Albert Frank	World War II
Schooler	Forrest Churchill	World War II
Scott	Richard D.	World War II
Sedgwick	John R	World War II
Shannon	Carl E.	World War II
Shaw	Jerome L.	World War II
Shubirg	George	World War II
Sines	Herbert E.	World War II
Smith	Harold Arthur	World War II
Smith	Lee I.	World War II
Smith	Ray E.	World War II
Smith	Robert L.	World War II

Above: The headstone of WWII veteran, Marvin Fleak in the Kirkersville Cemetery.

Below: Rows of WWI headstones in Cedar Hill Cemetery in Newark.

Above: Vietnam veteran, Kenn Merle Stanley, in Weaver Cemetery, near St. Louisville.

Above: Thomas E. Popham, WWI veteran, Eden Church cemetery near Purity.

Snelling	Norris	World War II
Spring	Bernard V.	World War II
St John	Wesley H.	World War II
Starrett	Charles Robert	World War II
Stevens	Earl M.	World War II
Stevens	Jr., William H.	World War II
Stricklin	Raymond E.	World War II
Swan	Robert E.	World War II
Swank	Glenn W.	World War II
Taafel	John W.	World War II
Tedesco	Michael T.	World War II
Thompson	Logan A.	World War II
Tiemann	James L.	World War II
Troendly	Donald P.	World War II
Turner	Gilbert E.	World War II
Tusing	Junior	World War II
Tyhurst	Dallas C.	World War II
Van Atta	Clark	World War II
Varner	Paul Richard	World War II
Wahl	Joseph L.	World War II
Walters	William	World War II
Watson	James Tod	World War II
Weiant, Jr.	Carl Andrew	World War II
Welsh	Forrest J.	World War II
White	Arthur T.	World War II
White	Glenn A.	World War II
Williamson	George Sidney	World War II
Winters	Charles R.	World War II
Wolfle	William R.	World War II
Woodbridge	John P.	World War II
Young	Clarence R.	World War II
Yurkschatt	Leonard G.	World War II
Zimmerman	George W.	World War II

World War I

Abbott	Thomas "Tod"		World War I
Andrews	Charles F.		World War I
Armstrong	Eldon Leroy		World War I
Arnold	Russell	Pataskala	World War I
Baughman	Henry C.	Johnstown	World War I
Beck	George	Newark	World War I
Beckholt	Edward C.	Utica	World War I
Berkshire	James E.	Newark	World War I

Bibler	Charles M.	Etna	World War I
Billman	Leonard		World War I
Bonham	Jesse F.		World War I
Brown	Samuel	Levenworth	World War I
Brown	William E.	Hebron	World War I
Buxton	Hugh L.		World War I
Campbell	Frank	Newark	World War I
Catton	Herbert		World War I
Connors	Eugene	Newark	World War I
Cooperrider	Loren H.	Glenford	World War I
Coyne	George A.	Newark	World War I
Crawford	Orville		World War I
Daugherty	Albert		World War I
Drake	George		World War I
Eswine	Russell C.		World War I
Evans	Homer	Croton	World War I
Fergus	Frank Charles	Newark	World War I
Frush	Walter Guy	Jack-town	World War I
Hannahs	Robert E.	Alexandria	World War I
Hayes	Raymond	Utica	World War I
Hite	Virgil O.	Granville	World War I
Hottinger	Homer		World War I
Hottinger	Zennie	Granville	World War I
Hursey	Ross	Gratiot	World War I
Hyatt	James Alden	Newark	World War I
Hyatt	James Allen	Johnstown	World War I
Jones	John K.	Newark	World War I
Jones	Kenneth		World War I
King	McKinley		World War I
Kinsely	Hugh	Utica	World War I
Koontz	Ralph G	St. Louisville	World War I
Korzenborn	Fred	Newark	World War I
Lamp	Ellis L.	Newark	World War I
Leedy	Arch		World War I
Lewis	Walter W.	Johnstown	World War I
Liber	Ernest T.	Newark	World War I
Livingston	Clyde D.	Newark	World War I
Lightner	Lawrence	Utica	World War I
McCurdy	William Joseph	Newark	World War I
McKnight			World War I
McWilliams	Clayton		World War I
Matheney	Enzley J.		World War I
Mathis	Wyvil L.	Johnstown	World War I
Mazey	Clark		World War I
Meckley	Homer C.	Newark	World War I

Above: The headstone of a Civil War veteran in Hanover Cemetery.

Below: Headstone of WWI veteran Clyde Hughes and his wife Myrtle, in Evans Cemetery near Saint Louisville. The unusual headstone is partially made of Flint Ridge flint.

Above: Veteran, Noble Swartz served as a PFC in the 324 Field Arty, 83 Div., in WWI. He lies in Jacksontown Cemetery.

Above: George Livingston served as a Pvt. in Co. A., 347 Machine Gun Bn. in WWI. He lies in Fallsburg Cemetery.

Below: Spanish-American War veteran, Charles C. Loar's headstone in the Hanover Cemetery.

Miller	Thomas O.	Newark	World War I
Moose	Roy J.	Outville	World War I
Morgan		Pataskala	World War I
Pickering	Channing	Johnstown	World War I
Pop	Isadore	Newark	World War I
Pound	Lee		World War I
Repede	Toma	Newark	World War I
Richardson	Forest	Wilkins Crn	World War I
Rollison	Dervey	St. Louisville	World War I
Roney	Dwight M.	Newark	World War I
Rostofer	Samuel		World War I
Scott	Orville G.	Utica	World War I
Smith	Charles Elmer	Newark	World War I
Starrett/Sterett	George	St. Louisville	World War I
Stricklen	Clarence		World War I
Sturman	Rollo B.		World War I
Swingle	Gray		World War I
Wheeler	James G.	Newark	World War I
Wheeler	Marion O.		World War I
Taylor	Harvey	Newark	World War I
Tucker	Gene	Newark	World War I
Varney	Irvin Scott		World War I
Wall	Lewis		World War I
Wheeler	James	Newark	World War I
Wheeler	Marion O.	Newark	World War I
White	Earl T.	Newark	World War I
Williams	Cecil R.	Newark	World War I
Williams	Thurman E.	Alexandria	World War I
Woolard	Wilmer A.	Buckeye Lake	World War I

Spanish-American War

No reliable list found.

Civil War

Abbott	George I.	Newark	Civil War
Ackley	Sterling		Civil War
Adams	Rufus B		Civil War
Adams	John Q.		Civil War
Alexander	Robert		Civil War
Allen	Morris		Civil War
Alward	Alfred		Civil War

Alward	George I.		Civil War
Anderson	Asbury		Civil War
Anderson	John S.		Civil War
Anderson	T.B.		Civil War
Armentrout	David E.		Civil War
Armstrong	David		Civil War
Atus	A.		Civil War
Bailey	Isaac	Hartford	Civil War
Bailey	Josiah		Civil War
Baird	Josiah		Civil War
Barnes	Thomas		Civil War
Barren	James		Civil War
Barrick	Lorenzo		Civil War
Barstow	G.H.		Civil War
Barstow	Judson		Civil War
Battee	A.J.		Civil War
Bauer	Jacob		Civil War
Bayles	George W.		Civil War
Beddoes	William		Civil War
Bell	Joseph		Civil War
Belt	John W.		Civil War
Benner	William A.		Civil War
Bennington	Nelson		Civil War
Berger	John C.		Civil War
Bird	Mervard		Civil War
Bishop	Henry		Civil War
Blade	Henry		Civil War
Bodle	John		Civil War
Bodle	William		Civil War
Bogle	Holmes		Civil War
Boughman	William		Civil War
Bounds	Perry		Civil War
Bowers	Andrew		Civil War
Brackett	Alonzo M.	Newark	Civil War
Brooke	Alanson S.		Civil War
Brookover	G.H.		Civil War
Brown	Israel		Civil War
Buck	W.H.		Civil War
Buckingham	Zadoc		Civil War
Burch	Leonidas		Civil War
Burrell	Theodore		Civil War
Buttricks	George		Civil War
Buxton	Rufus B.		Civil War
Camp	William		Civil War
Cane	Harvey		Civil War

Above: Beard-Green Cemetery at Dawes Arboretum.

Above: Revolutionary War veteran in Gaffield Cemetery at the intersection of Licking County Highways 40 & 25 near Outville.

Above: George Wise, Civil War veteran, Sanford Cemetery.

Above: Overgrown headstones in the old Bowling Green Cemetery near Marne.

Carter	John R	Civil War
Cary	Peter	Civil War
Case	W.E.	Civil War
Casteel	Henry	Civil War
Chambers	"Yank"	Civil War
Channell	Aaron N.	Civil War
Channell	Chester F.	Civil War
Channell	Joel D	Civil War
Chapin	James Sr.	Civil War
Chapin	James A.	Civil War
Cheek	John F.	Civil War
Clark	George	Civil War
Clark	J.R	Civil War
Clark	John F.	Civil War
Clemm	Harrison D.	Civil War
Cline	N.	Civil War
Clifton	Jonathan	Civil War
Coffman	Dever	Civil War
Colvin	William D.	Civil War
Condon	L.	Civil War
Conley	Michael	Civil War
Conrad	Clarence	Civil War
Cook	William J.	Civil War
Cooksey	William A.	Civil War
Cooksey	William B.	Civil War
Cooley	Lewis	Civil War
Coons	George	Civil War
Cooper	David H.	Civil War
Cornine	Harman	Civil War
Craft	William F.	Civil War
Cramer	W.H.H.	Civil War
Criswell	William	Civil War
Critchet	L.B	Civil War
Dare	John H.	Civil War
Davis	John S.	Civil War
Davis	Seymour	Civil War
Davis	Thomas J.	Civil War
Davis	Thomas O.	Civil War
Dill	John A.	Civil War
Dill	Thomas M.	Civil War
Dille	William	Civil War
Dispennett	Thomas	Civil War
Divan	Samuel	Civil War
Divine	David	Civil War
Dodson	David	Civil War

Donley	John	Civil War
Dorsey	Thomas	Civil War
Doyle	John	Civil War
Drake	John T.	Civil War
Drake	Silas	Civil War
Duckworth	Thomas L.	Civil War
Dunlap	John	Civil War
Dunn	David R.	Civil War
Dunn	Samuel H.	Civil War
Dunn	Thomas	Civil War
Eddington	Eleazer	Civil War
Edelblute	Theron	Civil War
Edwards	Alexander N.	Civil War
Ellis	Henry	Civil War
Ells	Charles A.	Civil War
Ensley	William	Civil War
Evans	Byron	Civil War
Evans	David	Civil War
Evans	Thomas	Civil War
Evans	Wesley	Civil War
Ewing	D.	Civil War
Finney	Walter E.	Civil War
Fleek	William H.	Civil War
Fletcher	William	Civil War
Flinn	Jerome	Civil War
Flowers	Harvey	Civil War
Follett	Dwight	Civil War
Fondersmith	David	Civil War
Foregrave	Michael	Civil War
Francis	John L.	Civil War
Francis	John R.	Civil War
Franks	Martin	Civil War
Frazier	William	Civil War
French	Ira B.	Civil War
Friddle	G.W.	Civil War
Frost	Ezra	Civil War
Frost	Joseph	Civil War
Frost	Levi	Civil War
Frost	J.W.	Civil War
Fry	Thomas Corwin	Civil War
Fulton	Henry	Civil War
Gants	Jacob	Civil War
Gardner	B.F.	Civil War
Gardner	John S.	Civil War
German	Henry A.	Civil War

Above: Jno. Wood, Civil War veteran, in the Old Colony Burying Ground, Granville.

Below: Two early veterans buried in the Beard-Green Cemetery at Dawes Arboretum.

Above: Headstone of Simon P. and John W. Holmes in Fallsburg Cemetery. According to the inscription on the stone, one or both of these men died at Andersonville Prison.

Above: A deteriorating gravestone tagged with a yellow label identifying it as a "U.S. Veteran." This grave is located in Mount Cavalry Catholic Cemetery along State Route 13 just south of Hopewell Drive in Newark, Ohio.

Gibboney	Silas A.	Civil War
Gibson	Joseph	Civil War
Gill	William S.	Civil War
Gillespie	Jasper L.	Civil War
Goff	Preston E.	Civil War
Gooding	Joseph W.	Civil War
Gorman	Timothy	Civil War
Gray	John	Civil War
Green	Albert G.	Civil War
Green	Benjamin L.	Civil War
Green	George	Civil War
Green	James W.	Civil War
Green	John W.	Civil War
Green	Joseph A.	Civil War
Gregory	Thomas	Civil War
Griffin	Michael	Civil War
Griffith	David	Civil War
Groves	Jacob	Civil War
Hagar	Jacob	Civil War
Haight	George E.	Civil War
Halliday	Franklin	Civil War
Hamilton	Arthur T.	Civil War
Hamilton	Henry	Civil War
Hamiliton	Robert J.	Civil War
Hamiliton	William M.	Civil War
Hand	James G.	Civil War
Handley	Joel D.	Civil War
Hanly		Civil War
Hanley	James	Civil War
Hardesty	Nelson	Civil War
Harris		Civil War
Hartzell	Hiram	Civil War
Harvey	Charles	Civil War
Haslop	Guilford D.	Civil War
Hatch	Osmer	Civil War
Hatfield	J.W.	Civil War
Haughey	Frank A.	Civil War
Hawke	Elijah	Civil War
Hayner	B.	Civil War
Hayner	Enoch	Civil War
Henne	Michael	Civil War
Henry	John	Civil War
Hermon	R.	Civil War
Hickey	Henry	Civil War
Hickey	Win.	Civil War

Hill	Benjamin	Civil War
Hill	Caton	Civil War
Hill	Ezra	Civil War
Hill	John W.	Civil War
Hillery	W.	Civil War
Hines	William	Civil War
Hirst	Thomas A.	Civil War
Hollar	David A.	Civil War
Holland	John	Civil War
Holler	Hiram	Civil War
Hollinger	Jacob B.	Civil War
Holmes	Wesley	Civil War
Holtsberry	Isaac	Civil War
Holtsberry	Thomas M.	Civil War
Honnegger	Jacob	Civil War
Hoover	George M	Civil War
Hopkins	Ebenezer W.	Civil War
Hopkins	Lewis	Civil War
Howell	Enos	Civil War
Howell	James W.	Civil War
Huff	Franklin A.	Civil War
Hughes	Alexander	Civil War
Hughes	Elias	Civil War
Hughes	J.	Civil War
Hughes	Joshua	Civil War
Hughes	Richard	Civil War
Hutchins	G.W.	Civil War
Hynus	Joseph	Civil War
Ihbotson	Henry	Civil War
Irwin	George	Civil War
Irwin	Henry	Civil War
Irwin	John S.	Civil War
James	John	Civil War
Jennings	Joseph W.	Civil War
Jenkins	Eugene J.	Civil War
Jerret	George	Civil War
Johns	George	Civil War
Johnson	Thomas	Civil War
Jones	Evan	Civil War
Jones	George W.	Civil War
Jones	John	Civil War
Jones	John H.	Civil War
Jones	S.D.	Civil War
Jones	Thomas	Civil War
Kent	Henry	Civil War

Above: Headstone of Civil War casualty, John Elias Hughes in the Calvanistic Welsh Church Cemetery on Sharon Valley Rd.

Below: Marker for Rev. John Emmet, veteran of the American Revolution, in North Lawn Cemetery, Utica.

Above: Pvt. Edelbert H. Cooley served during the Civil War and died at age 16. He is buried in the Alexandria Cemetery.

Below: Rows of headstones in the old cemetery at Claylick.

Kindle	Wesley	Civil War
Kindred	John	Civil War
King	Lewis	Civil War
Kissel	David	Civil War
Kite	Adam	Civil War
Knapp	Charles E.	Civil War
Kugler	Joseph	Civil War
Kumpf	George	Civil War
Lake	Sylvanus A	Civil War
Lake	Vincent	Civil War
Lalley	James	Civil War
Lane	William H.	Civil War
Laner	John	Civil War
Larrabee	John W.	Civil War
Larmbee	William H.	Civil War
Latham	John	Civil War
Layman	Daniel D.	Civil War
Leese	Joseph A.	Civil War
Lemert	John A.	Civil War
Lemert	Nathan F.	Civil War
Lemert	Thaddeus	Civil War
Lemert	Thomas J.	Civil War
Levingston	Andrew	Civil War
Levingston	William	Civil War
Lewis	Nelson	Civil War
Licer	Patrick	Civil War
Lickliter	Henry	Civil War
Lincoln	Lyman H.	Civil War
Little	John W.	Civil War
Looker	Hector	Civil War
Loughman	Robert H.	Civil War
Loveland	Charles H.	Civil War
Lowry	John	Civil War
Loyd	David	Civil War
Luther	Charles	Civil War
Lyman	Joseph	Civil War
Lyon	Matthew	Civil War
Maddox	James	Civil War
Mann	Macey	Civil War
March	Ira	Civil War
Mark	J.	Civil War
Martin	M.	Civil War
Marvin		Civil War
Mason	William C.	Civil War
Matthews	Delano	Civil War

McBride	John	Civil War
McCloy	James	Civil War
McComb	Uriah A.	Civil War
McCrum	Robert	Civil War
McDougal	Leonidas	Civil War
McFarland	Benjamin	Civil War
McGaffick	Joseph	Civil War
McKnight	I.	Civil War
McKnight	John W.	Civil War
McMillen	Charles	Civil War
McVicker	George W.	Civil War
Meekens	David	Civil War
Meekens	Richard B.	Civil War
Merchant	Dallad	Civil War
Meredith	Benjamin D.	Civil War
Merrill	John Quincy	Civil War
Messenger	Madison C.	Civil War
Miles	Stephen	Civil War
Miller	John	Civil War
Miller	John R.	Civil War
Miller	D.H.	Civil War
Miller	Samuel	Civil War
Mimms	George	Civil War
Minton	Isaac S.	Civil War
Minton	John N.	Civil War
Minton	William J.	Civil War
Mitchell	Hiram P.	Civil War
Moore	D.A.B.	Civil War
Moore	James	Civil War
Moore	John	Civil War
Moore	Peter A J.	Civil War
Moore	Robert A.	Civil War
Moorehead	John M.	Civil War
Morris	E.J.	Civil War
Morris	William H.	Civil War
Mossman	James Y.	Civil War
Munson	Francis	Civil War
Munson	G. Adolphus	Civil War
Murphy	Wesley	Civil War
Murrel	John L.	Civil War
Myers	Jacob	Civil War
Myers	Loyd H	Civil War
Nelson	Henry	Civil War
Newman	James	Civil War
Newman	William	Civil War

Above: Entrance to Hollar Cemetery along State Rt. 13 just north of Newark, Ohio. White headstone in center memorializes veteran John Sparks, an early Newark-area resident who enlisted in the Army in 1804 and accompanied Zebulon Pike on his two famous expeditions.

Below: Marker for John O. Hupp, veteran of WWII, in the Smith Chapel Cemetery.

Above: Civil War headstone in Sanford Cemetery, Rt. 13 south of Newark.

Below: Mexican War veteran in the main cemetery at Johnstown.

Nichols	John M.	Civil War
Nichols	Townsend	Civil War
Nolan	John A.	Civil War
Nolan	William	Civil War
Norman	John L.	Civil War
Ogilvie	Stewart J.	Civil War
O'Harra	James	Civil War
Oliver	William	Civil War
Ormsby	D.C.	Civil War
O'Sullivan	Patrick	Civil War
O'Sullivan	Timothy	Civil War
Oster	Charles	Civil War
Outcalt	R.V.	Civil War
Overholt	John S.	Civil War
Page	Hiram	Civil War
Palmer	Alonzo B.	Civil War
Palmer	Samuel	Civil War
Parker	Alfred	Civil War
Parker	Burns T.	Civil War
Parr	Marquis D. C.	Civil War
Parsons	Hiram H.	Civil War
Parsons	W.D.	Civil War
Patterson	David	Civil War
Patton	John S.	Civil War
Pence	Isaac	Civil War
Philbrook		Civil War
Pierce	Theodore W.	Civil War
Pierson	Samuel W.	Civil War
Pierson	Walter	Civil War
Pittsford	Timothy H.	Civil War
Poland	Wesley	Civil War
Poor	A.M.	Civil War
Porter	Amos A.	Civil War
Porter	William	Civil War
Potter	William	Civil War
Pratt	Lyman	Civil War
Preston	James N.	Civil War
Preston	William O.	Civil War
Price	Clinton	Civil War
Price	Jonathan	Civil War
Price	Joseph	Civil War
Price	Joshua	Civil War
Price	Samuel	Civil War
Priest	Joseph	Civil War
Priest	Solomon	Civil War

Prior	Anthony	Civil War
Proctor	David	Civil War
Prudent	Henry C.	Civil War
Pumphrey	Albert	Civil War
Ramey	Elijah	Civil War
Reader	George T.	Civil War
Readhead	Benjamin	Civil War
Rechell	John	Civil War
Reed	George B.	Civil War
Rees	Richard	Civil War
Redhead	Joseph	Civil War
Redmond	Sylvester	Civil War
Redmond	William	Civil War
Remington	Nelson	Civil War
Reynolds	Horace	Civil War
Richards	Samuel	Civil War
Ridenour	William	Civil War
Rider	William	Civil War
Riefff	John	Civil War
Roberts	Charles	Civil War
Roberts	Leroy	Civil War
Robinson	Alonzo	Civil War
Robertson	John	Civil War
Root	Martin	Civil War
Rose	Albert	Civil War
Rose	Daniel	Civil War
Rose	Henry D.	Civil War
Rose	Samuel L.	Civil War
Rudolph	Jacob	Civil War
Ryland	John	Civil War
Runnels	Stephen W.	Civil War
Runnion	Joseph	Civil War
Rush	George C.	Civil War
Sandals	William H.	Civil War
Savory	Charles	Civil War
Sawyer	Adam	Civil War
Sawyer	Samuel C.	Civil War
Sayers	Aaron	Civil War
Scott	William	Civil War
Selbey	Byron	Civil War
Sessor	Thomas	Civil War
Setzer	Philip W.	Civil War
Seymour	Henry S.	Civil War
Shambaugh	J.B.	Civil War
Shaw	Andrew J.	Civil War

Above: The headstone of Civil War veteran, Sergt. John McDowell, Co. F, 135 Ohio Inf., in the cemetery at Wilkins Corner.

Below: The grave of Methias T. Smoots, Civil War veteran, in Utica's North Lawn Cemetery.

Abram P. Westbrook, the oldest veteran of the War of 1812, died in Newark on October 10, 1880. According to his obituary, he was born in 1778, fought in the War of 1812, and died at the age of "one hundred and two years one month and eight days old." He is buried in Cedar Hill Cemetery.

Source: *The History of Licking County O.* by N. N. Hill, Jr., 1881.

Below: A veteran of the Spanish-American War in Cedar Hill Cemetery, Newark.

Shepherd	Nelson M.	Civil War
Sherman	Reuben	Civil War
Shipps	Newton S.	Civil War
Shobble	Daniel	Civil War
Shohoney	Edward	Civil War
Shutt	John W.	Civil War
Simpson	Merideth	Civil War
Sinnett	Rcuben	Civil War
Sinsabaugh	Cyrus	Civil War
Smart	Charles A.	Civil War
Smart	George	Civil War
Smart	Joseph A.	Civil War
Smith	George M	Civil War
Smith	Hiram W.	Civil War
Smith	John	Civil War
Smith	John W.	Civil War
Smith	Sidney	Civil War
Smith	Wesley V.	Civil War
Somerville	Levi	Civil War
Spellman	Abraham	Civil War
Sperry	Joseph	Civil War
Stafford	Charles W.	Civil War
Steel	George	Civil War
Stevens	Ira W.	Civil War
Stevens	John D. D	Civil War
Stevens	Lewis L.	Civil War
Stevenson	George W.	Civil War
Stockton	Thomas M.	Civil War
Stout	Felix	Civil War
Streeper	Frank W.	Civil War
Strother	Benjamin	Civil War
Sullivan	John	Civil War
Sullivan	Patrick	Civil War
Talbott	Rufus H.	Civil War
Tarr	George F.A.	Civil War
Taverner	John	Civil War
Taylor	James W.	Civil War
Taylor	William	Civil War
Teagarden	John	Civil War
Tharp	William	Civil War
Thomas	Samuel	Civil War
Thomas	Thomas	Civil War
Thorn	George F.A.	Civil War
Thrall	Aaron	Civil War
Thurston	Elmore M	Civil War

Tobin .... Civil War
Troy .... Morris .... Civil War
Towne .... James .... Civil War
Tucker .... Jesse H. .... Civil War
Turner .... Lyman D. .... Civil War
Twining .... Cyrus .... Civil War
Upham .... G. Baxter .... Civil War
Van Allen ..... John .... Civil War
VanHorn ...... Arthur J. .... Civil War
Van Kirk ...... George W. .... Civil War
Vermillion ... Richard .... Civil War
Vinning .... William H.H. .... Civil War
Waggoner .... John .... Civil War
Walker .... Charles .... Civil War
Wall .... Simeon B. .... Civil War
Waller .... George .... Civil War
Walsh .... Thomas .... Civil War
Ward .... Charles V. .... Civil War
Ward .... Daniel .... Civil War
Ward .... David .... Civil War
Warden .... John F. .... Civil War
Warner .... John .... Civil War
Webb .... Hiram .... Civil War
Weaver .... John .... Civil War
Webber .... Jesse C. .... Civil War
Weekly .... David T. .... Civil War
Wells .... Henry A. .... Civil War
Wells .... G.C. .... Civil War
West .... John T. .... Civil War
Westbrook ... R.S. .... Civil War
Weyrearch ... Thomas J. .... Civil War
Wharton .... Arthur .... Civil War
Wharton .... George .... Civil War
Wheelock .... James L. .... Civil War
White .... James .... Civil War
Whitehead ... George F. .... Civil War
Whitehead ... S.P. .... Civil War
Wickham ..... J.S. .... Civil War
Willey .... William J. .... Civil War
Wilcox .... Benjamin .... Civil War
Wilcox .... Marvin .... Civil War
Williams ...... Amos P. .... Civil War
Williams ...... James .... Civil War
Williams ...... Joshua .... Civil War
Williams ...... Lewis .... Civil War

Above: The old cemetery in Johnstown, Ohio.

Below: Tree-shaped monument in Hanover Cemetery.

Above: The ornate headstone of Seth Mead(?), a veteran of the War of 1812, in the Homer Cemetery.

Below: The grave of E. Warren, veteran of the Mexican War, in Granville's Old Colony Burying Ground.

Williams	William	Civil War
Williamson	Basil	Civil War
Wilson	Jacob S	Civil War
Wilson	Philip	Civil War
Willyard	John	Civil War
Winters	William H.	Civil War
Wise	John A	Civil War
Wolf	Peter	Civil War
Wood	Sidney R.	Civil War
Woodcock	George W.	Civil War
Woodruff	J.M.	Civil War
Wright	Collin C.	Civil War
Wright	George E.	Civil War
Wright	P.	Civil War
Zipperer	Jacob	Civil War

Mexican War

Wilkins	Robert	Mexican War
Courson	Harvey	Mexican War
Lansing	William	Mexican War

Indian Wars

No reliable list found.

War of 1812

No reliable list found.

Revolutionary War

No reliable list found.

ANDERSON, Eric B., (editor), "Licking County Historical Review, 175 Years." Published by The Advocate, 1994.

BRADLEY, George W., III. "From 'Missile Base' to 'Gold Watch:' An Illustrated History of The Aerospace Guidance and Metrology Center and Newark Air Force Station." Office of History, Aerospace Guidance and Metrology Center, Newark Air Force, Ohio, 1982.

BRISTER, E. M. P., "Centennial History of the City of Newark and Licking County Ohio." Published by The S.J. Clarke Publishing Co., 1909.

BROCKWAY, Chance, "Images of America – Newark." Published by Arcadia Publishing, 2004.

FLEMING, Dan and THARP, Robert, "A Bicentennial History of Licking County, Ohio 1808-2008." Published for the History Committee of the Licking County Bicentennial Commission, Inc., 2008.

HARTROFT, PHYLLIS M. and KOHSER Nina C., "Licking County History" As told in a decade of feature articles from The Licking County Historical Society's Quarterly. Compiled and edited, 2002.

HILL, N. N., Jr., "History of Licking County O., It's Past and Present."

Published by A. A. Graham & Co., Publishers, 1881.

HOWE, Henry, LL. D., "Historical Collections of Ohio in Two Volumes." Published by The State of Ohio, 1896.

HUFF, W. T., "Newark's Pioneer Graveyard (A Brief History of the Sixth Street Cemetery)." Published by W. T. Huff, 1999.

HUFF, W. T., "Pioneer Life in Licking County – Selected Works of Ben Jones (1865-1947)." Published by The Newark Bicentennial Commission and The Advocate, 2002.

KEESEE, Dennis M., "Too Young to Die: Boy Soldiers of the Union Army 1861-1865," published by Blue Acorn Press, Huntington West Virginia, 2001.

KEIRNS, Aaron J., "Black Hand Gorge, A Journey Through Time." Published by Little River Publishing, 1995.

LAWYER, J. P., Jr. "History of Ohio: From the Glacial Period to the Present Time." Published by Union Publishing Co., Columbus, Ohio, 1904.

LISSKA, Anthony J. , WILHELM, Clarke L., (editors) and others, "Granville, Ohio – A Study in Continuity and Change." Published by Granville Historical Society and Denison University Press, 2004.

RUTTER, Connie L. and GARTNER, Sondra Brockway, "Images of America – Licking County." Published by Arcadia Publishing, 2007.

WILKIN, Janice M. "Heath, Ohio: Hometown Memories." City of Heath, Ohio, 2001.

Aaron J. Keirns

Aaron is the author of several books on Ohio history topics. He holds a B.A. in Anthropology from The Ohio State University and is a professional writer, designer, publishing consultant and public speaker. He serves as Director of Communications at The Works Museum in Newark.

Aaron is a Vietnam-era veteran, drafted in Licking County in 1969. Most of his time in service was spent in the U.S. Army's 82nd Airborne Division, 2/504th Infantry at Fort Bragg, North Carolina. He was honorably discharged in 1972 at the rank of E-5. He lives in Knox County with his wife Bernice.

Nathan J. Keirns

Nathan is a native of Ohio with a long-standing interest in local history. He holds a Master of Arts in Sociology, teaches undergraduate courses, and is employed by a social service agency in Licking County.

Nathan is committed to serving the community, is currently a member of the Licking County Leadership Program Class of 2010, and has served on the Services to Armed Forces Committee with the American Red Cross of Licking County. Nathan lives in a small town with his wife, Amber.

Made in the USA
Monee, IL
04 November 2026

67377031R00060